Hartmut Wittig

Intelligent Media Agents

W0091603

Multimedia Engineering
hrsg. von Wolfgang Effelsberg und Ralf Steinmetz

Die multimediale Revolution ist in vollem Gange. Neuere Arbeitsplatz-rechner und viele PCs, die am Markt erscheinen, haben heute schon Audio-Komponenten eingebaut, und in zunehmendem Maße findet man auch Hardware- und Softwareunterstützung für die Darstellung von Bewegt-bildsequenzen. Die multimediale Art der Interaktion mit dem Computer ist viel effizienter und benutzerfreundlicher als die Interaktion über die Ein- und Ausgabe von Texten und hat deshalb ein hohes Zukunftspotential. Zugleich eröffnen die Techniken der computergestützten Kooperation neue Möglichkeiten zur Teamarbeit in vernetzten Unternehmen.

Ziel der Reihe ist es, den Leser über Grundlagen und Anwendungen der Multimedia-Technik und der Telekooperation zu informieren. Die Reihe umfaßt Lehrbücher, einführende und umfassende Standardwerke sowie speziellere Monographien zu den Themen Multimedia, Hypermedia und computergestützte Kooperation. Es geht dabei beispielsweise um Fragen aus den Bereichen Betriebssysteme, Rechnernetze, Kompressionsverfah-ren und grafische Oberflächen. In der Art der Darstellung wendet sie sich an Informatiker und Ingenieure, an Wissenschaftler, Studenten und Prak-tiker, die sich über dieses faszinierende interdisziplinäre Thema informie-ren wollen.

Bisher erschienen:

Synchronisation in kooperativen Systemen
von Erwin Mayer

Multimediale Kiosksysteme
von Wieland Holfelder

Bildkompression mit Fraktalen
von Michael F. Barnsley und Lyman P. Hurd

Multimedia, Hypertext und Internet
von Jakob Nielsen

Entwicklung verteilter Multimedia-Applikationen
von Thomas Käppner

Sicherheit für Videodaten
von Thomas Kunkelmann

Intelligent Media Agents
von Hartmut Wittig

Vieweg

Hartmut Wittig

Intelligent Media Agents

Key technology for Interactive Television,
Multimedia and Internet Applications

vieweg

1st Edition 1999

http://www.vieweg.de

Cover design: Ulrike Weigel, www.CorporateDesignGroup.de
Printing and binding: Lengericher Druckerei Hubert & Co., Göttingen
Printed on acid-free paper
Printed in Germany

ISBN 3-528-05706-8

Acknowledgments

I want to thank IBM, Multimedia Software GmbH and Deutsche Telekom Computer Service Management GmbH for providing the innovative environment which inspired me to start and complete this book about intelligent agents.

When I wrote the book, I was supported by many people. My discussion with Nicholas Negroponte, Ralf Steinmetz and Winfried Kalfa initiated this effort by supporting my vision that internet, interactive television, digital media and intelligent agent technologies will closely be integrated in the next century.

The strongest influence were the contibutions and joint work with my former colleagues Michael Ehrmanntraut, Keith Hall, Carsten Griwodz, Sharyar Vazirian, Stefan Michel and the IBM and Multimedia Software GmbH Dresden teams. Thank you very much.

The most important persons I left to thank are my wife Ulrike and my family. They gave me the opportunity to finish the last 10% of the book in 90% of the time.

Hartmut Wittig

Darmstadt, March 1999

Hartmut.Wittig@telekom.de

Preface

Telematic services have initiated an economic and social revolution which will deeply affect our lives in the future. High-speed broadband networks and network computing are basic elements of the information management platform of tomorrow. One of the main challenges is to provide a variety of information services to the user which are tailored to individual needs. More intelligent and individual services are therefore required to master the glut of information.

Agent technology has become one of the fastest growing areas of the Internet. This boom is driven by users who are looking for more intelligent solutions for their problems in interactive networked applications.

The book written by Hartmut Wittig provides a description of intelligent agent, multimedia architectures, and software technologies. It covers innovative technologies for designing and implementing future key applications in the Internet, intranets and extranets, applications which will help private and business users to access and use information in a very productive and efficient way. Hartmut Wittig has combined his excellent practical experience with his theoretical knowledge in agent technology, multimedia systems and media content. Readers will find this book very attractive because it highlights various research and development trends and provides an exciting insight into a world rich in telematic services.

Hagen Hultzsch

Member of the Board

Deutsche Telekom AG

Table of Contents

List of Figures

List of Tables

1 Introduction

Television

Television was born when the German student Paul Nipkow used two perforated wheels, a selen cell, a lamp and cable to spend Christmas with his mother even though she was kilometres away from him.

He invented the electrical transmission of moving pictures. Since then television has become increasingly popular, and today television is the preferred choice of leisure activity. Every second household in the world is equipped with at least one television set, and more than 80 percent of the people in these households use it every day.

Multimedia Services

The term "Multimedia" describes the integration of various media types (e.g., audio, video, two-dimensional and three-dimensional graphics, text) into one single information format. In addition to multimedia capabilities, the term "interactive television" (iTV) stands for a set of interactive services applied to television providing:

- interactive selection capabilities such as navigation in branched movies
- access to television programs, i.e., video-on-demand
- access to multimedia data associated with the television program

With the advent of iTV, new applications such as games, pay-per-view, movies or karaoke-on-demand have also been introduced. Prototypical interactive television systems are now available.

Digital transmission technologies and data compression methods provide the capability to send and receive hundreds of programs. This capability can be used to produce and transmit more TV programs and channels than are currently being offered including single-interest channels, program bundles and international programs.

Problems

The technical knowhow and infrastructure already exists to transmit these hundreds of TV channels. Satellite, cable and terrestrial networks deliver the requisite capacity for such transmissions. The problem arises when traditional user navigation is applied: Today, the TV viewer selects program by pressing TV remote control buttons. Assuming hundreds of TV channels are available, this navigation method is a problem for the television users. The vest number of programs being offered precludes the possibility of obtaining any overview of theme in a time. Assuming the availability of hundreds of TV channels, the traditional method of navigation will prove unworkable. In the worst case scenario, the user is unable to locate an appropriate television program or needs to first search for a very long time.

Therefore, it is necessary that new navigation systems be developed. Having evaluated user requirements, a more intelligent approach to improve the quality of navigation is proposed. Intelligent agents play a defining role in intelligent navigation systems.

An introduction into the recent research results in the area of television and selected examples of innovative television applications are given in the subsequent sections.

1.1 Research in Interactive Television Systems

In the recent years, research in the area of interactive TV systems has been aimed at creating technologies that improve functionality, quality, compatibility, and usability.

Functionality

Existing television systems do not have an integrated feedback channel. The provision of an integrated feedback channel without blocking the existing communication services is one of the functional goals of the network related research in interactive television systems (see [Steinmetz 96]).

A feedback channel is essential to implement interactive applications such as home shopping, video-on-demand, video conferencing, pay-per-view, or user voting.

Quality of Service

Digital and interactive television systems must provide a good quality of service (e.g., in terms of availability, robustness, video resolution and frequency). Several systems delivering a high quality of service for computer operating systems and networks have been developed and analysed (e.g., [Wolf 95], [Zhang 93]). Technologies for new high speed networks (e.g., ATM and Q.2931 quality of service interface) and existing computer and telecommunication networks allow for the delivery of high-quality multimedia data. As a result of this research, new digital transmission schemes and resource management facilities can guarantee quality services and a continuous scaling of the media quality and bandwidth.

Compatibility

Compatibility is essential to permit the use and distribution of various kinds of contents and interoperability within an open distributed system and networking environment.

Using standards can help reduce the overall costs of application programming and maintenance by allowing the development of shared tools and shared communication components. Standardization in the area of interactive television systems is done by the following international and industrial standardization organizations:

- International Standardization Organization
- Digitial Audio-Visual Council
- International Multimedia Association

For example, the standardization of media formats in interactive television systems includes images (e.g., JPEG [ISO 10918], JBIG [ISO 11544]), audio and video (MPEG-1 [ISO 11172], MPEG-2, Intel Indeo), synchronization information (MPEG-2 program and transport streams), and portable applications (MHEG-5 [ISO 135221]).

Usability

Today, television usability is very high given the simple user interface which even the most unskilled users can operate. In interactive television systems additional navigation services are required to take advantage of the enhanced set of functions without impairing the system's usability. Compared to traditional television systems, interactive TV systems additionally provide service interactivity and an extended functionality.

As a result of these extensions, the user interface seem to be more complex. Maintaining usability in an environment of increased functionality can be seen as a key design goal. Therefore, interactive television applications must be developed which take into account that users possess varying degrees of technical competence.

A major group of users is completely unfamiliar with digital interfaces and expects to have television interfaces as simple to operate as they are today. Due to the complex information structure in interactive television systems, the process of program and information search is expected to be more complex than before. A more intelligent approach which personalizes the handling of the user interface to suppress functionality too complex for a given individual user.

1.2   Applications

Interactive television systems contain a variety of applications offering public multimedia services to the user:

- television and interactive television
- information services
- home shopping
- games

As the prerequisite to personalization, differences between traditional services and iTV services are analysed below.

1.2.1   Television and Interactive Television

There are three main directions in interactive television development which can be seen when observing the worldwide development of prototypes and their field trials:

(1) The digital coding is replacing the analog one in media transmission.

(2) The unidirectional communication from the content provider over the distribution network managed by network providers to the curbs and homes (e.g., FTTC - fibre to the curb and FTTH - fibre to the home) will be replaced by bidirectional communication using back-channel networks to carry signals in the reverse direction.

(3) The variety of movie types will expand from the current mode of invariant linear delivery to include first branched and then variable movie structures. Whereas linear movies deliver an invariant sequence of frames and plot developments, branched movies offers the user the opportunity to interactively direct the subsequent directions of a film by selecting one branch from many thus determining the film directions.

For example, this user participatory mode can also enable users to interactively cast a vote for their number one piece in a music show. Figure 1 illustrates these projected developments in interactive television systems:

Today, linear programs are broadcast via unidirectional analog communication channels. In the first interactive television field trials begun in 1995, three modes of television were tested:

(1) In the digital video broadcasting mode and (2) the Near-Video-on-Demand mode a number of television channels is digitally broadcasted without having feedback channels. In the (3) video-on-demand mode interactivity is introduced by dedicated

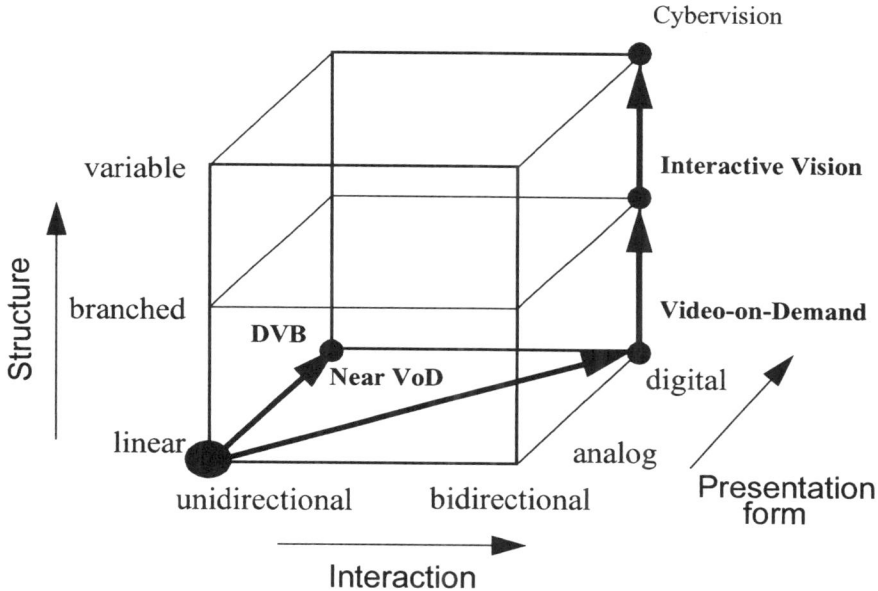

Figure 1: Trends in Television Technology

back-channel networks. Testbeds for the above modes of television have extensively been used (e.g., Deutsche Telekom trial in Nuremberg [MMS 96a]), followed by futher trials such as DVB over satellites, or WebTV networks).

Additional modes can be defined for structured movies. They have been called "interactive vision" and "cybervision". In the interactive vision mode, movies can be substructured to include decision frames in which the user can select from a number of potential plot developments to determine how the film will unfold.

The cybervision mode implements virtual environments and virtual reality (VR) in an interactive television environment. As a prerequisite, the TV set-top boxes must support such VR equipment as a head tracking system, and data gloves. In the cybervision mode the structure of the movie is continuously variable. The script of a cybervision movie does not contain a description of acting itself, but rather a description of the virtual environment setting. Additionally, a set of rules defines a

5

virtual actor's reaction to a viewer's action. It is not clear whether it is realistic to expect the interactive vision and cybervision modes in the near future. Therefore, the approach presented in this book is restricted to the digital video broadcasting and video-on-demand scenarios.

Digital Video Broadcasting (DVB) / Near-Video-on-Demand (Near VoD)

Digitally broadcasted television is an asymmetric service which provides for the transfer of digitally encoded TV channels to the end-users. DVB is based on a rigid time structure: a rewinding of the program is not possible. Near-Video-on-Demand (NVoD) has also been implemented in iTV field trials.

NVoD is the broadcast of one program over parallel channels at staggered start times. For example, the transmission over the second channel starts 10 minutes after the first transmission. This staggering enables forwarding and rewinding the program in fixed intervals by selecting broadcast channels. As in DVB, a user feedback channel is not provided.

Video-on-Demand

Video-on-Demand is defined by the ATM forum as:

> "an asymmetrical unidirectional service which provides the transfer of digitally compressed and encoded video information from a source, typically a video server, to a destination, typically a set top terminal (STT). At the decoder, the streams are reassembled, decoded, digital-to-analog converted and presented at the monitor."

Video-on-Demand is receiver-initiated and does not depend on a fixed time structure. VoD extends DVB services by providing facilities for digital video recording and storing of digitally broadcasted programs in a service provider archive. When a user orders a video, the request is sent from the set-top box over the back-channel network to the video server, and is evaluated by the controller of the server. A real-time network connection is then set up between the server and the set-top box, the video being sent over the distribution networks to the end-station.

Media Pools

A media pool is a system in which VoD and DVB modes are mixed as shown in Figure 2. The identity of traditional TV service providers in the broadcast mode is the same as in the traditional analog broadcast mode of television. Today, the media offerings in the video-on-demand mode are dominated by movies. There are possibilities to form the corporate identity for service providers and to distribute their programs: In the future, the entire programs which are distributed over the broadcast channels will also be offered in a video-on-demand mode. This can be achieved by copying the media content from an internal archive to a publicly accessible media server.

Television service providers can thereby also implement media pools which contain future programming or which are bundles of a set of special interest channels produced by one content provider. Users can alternatively choose either to use the traditional time slotted mode of DVB, or to request the program from these media pools.

In interactive TV systems comprised of both DVB and VoD, the programming on the distribution channel and the media pool contents will form the corporate identity of the content provider. Advantages for both content providers and users, media pools better reflect the specific interests of the viewers by offering more than the 24 hour of programming daily. By applying specific filters to the broad selection of programs offered in a traditional media pool, users can customize their programming.

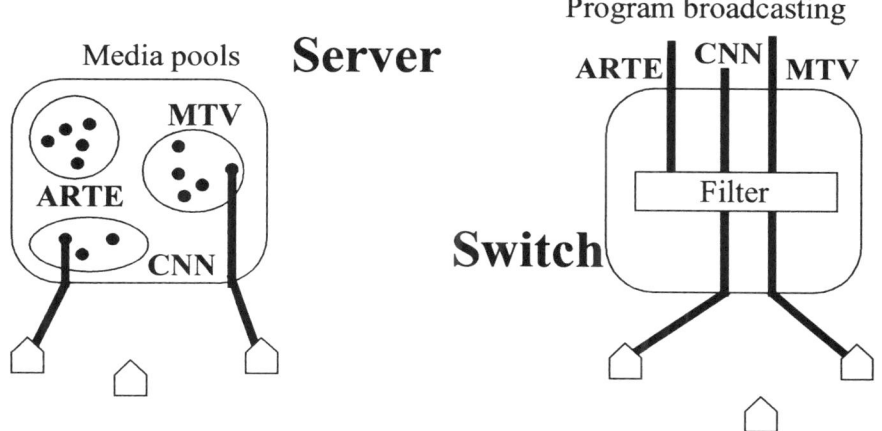

Figure 2: Media Pools and Program Broadcasting

Research Projects and Field Trials

It is expected that both VoD and DVB modes will coexist in the interactive television system environment. Digital video broadcasting and video-on-demand are being implemented and tested. In 1994-99 several research projects for interactive services were realized, e.g., the German Globally Accessible Services (GLASS)[1] project. In addition to research projects, there are field trial test beds involving 50 to 4000 test users. The main objective was to analyse various first-mile networks and multimedia application services (e.g., television [MMS 96a] or tele-shopping [MMS 96b]). Table 1 contains example parameters of field trials in Germany.

Location	Berlin	Köln	Nürnberg	Darmstadt
Network Technology	ISDN, Fiber Cable	ADSL	ADSL	ADSL
Area	Down-town	Down-town	Greater Town	Greater Town
Start Date	1994	1995	1995	1996
Duration (years)	1	4	3	1
Services	Services on Demand, Tele-Shopping, NVoD, VoD, Pay-per-view, Pay-per-channel			

Table 1: Field Trials in Germany

1.2.2   Value-Added Services

In interactive television a wide range of multimedia services is offered including tele-shopping, information retrieval, and multimedia communication with such features as multimedia mail and video conferencing.

---

1. GLASS has been initiated by DeTeBerkom. The system has been implemented within a consortium of companies (DEC, Grundig, and IBM ENC), research institutes (GMD FOCUS) and universities (TU Berlin). The author participated in the project team in 1994 and 1995.

Tele-Shopping

The first field trials have included shopping applications which enable users to buy or rent wares from virtual warehouses in the iTV system (see [MMS 96b]). There are two kinds of tele-shopping applications: The first involves sending an advertisement during a television program. At that time several shopping pages are loaded in the background.By activating specific hot-keys located on the remote-control order pages are called-up on the screen. After order forms have been completed the user can electronically submit them to the store by pressing the send-button. The second shopping application consists of special presentations developed by mail-shipping companies to offer an iTV screen display of their catalogs. Users can navigate through displays of the articles, consider advertisement clips and test reports, as well as price and discount information. Electronic shopping permit flexible cost management strategies, allowing prices to reflect circumstances. For example, users who place their order in an off-peak hour could be given a discount, or the price of an item could automatically rise to reflect its scarity.

Information Retrieval

Information can be retrieved by watching broadcast news television at pre-defined times or by accessing information databases on-line. The back-channel allows the user to specify his demands for information, and to selectively view news matching his interests. For example, users can get regional (e.g., district-wide, country-wide) news channels, or subscribe to digital news magazines. Older news can be stored in archives and be retrieved by users on demand. The interactive television system can also be used for receiving business related data, e.g., real-time stock market data, or news tickers. A very popular information retrieval application is access to the worldwide web. The information retrieval type can be compared with news-on-demand applications over the internet. Similar to content-based news filtering (e.g., the Hynode project [Griwodz 97]) users are able to specify preferences. News are filtered according to individual user preferences.

Tele Communication

Electronic mail and telephony as the digital and analogous services of telecommunication networks are expected to become additional services offered in an iTV application environment. Depending on the capacity of the back-channel, videos can also be recorded in real-time in the home and transmitted from homes to other users, e.g., in video telephony or a conferencing scenario. Together with television, videophony allows users to remotely participate in the program, for example in quiz or game shows. Comments and suggestions can be sent to the television studios in real-time, or electronic votes can be sent by using electronic mail programs.

1.3   The Problem of an Information Glut

In the last decade, the number of information appearance (e.g., via radio, television, Internet) has grown tremendously. This development can be observed in almost every mode of iTV systems. The only strategy to control this flood of information is to filter it. Individuals vary in their interests and needs, thus information filtering strategies which seek to meet these individual profiles will differ. Each information filtering strategy is based on the personal knowledge and experience of the individual user.

As described, digital data compression technologies and global broadband networks will be brought into traditional television systems. More information will be transmitted by using digital satellite and cable networks. For example, in the digital TV satellite system ASTRA more than 100 channels are provided (e.g., used by the digital television pool DF1). Future navigation systems should take into consideration that there is more information being sent and additional flexibility being offered by the providers of programs and interactive multimedia services.

The fundamental assumption of this book is that using navigation systems similar to those employed today will overwhelm traditional users with information, resulting in problems when attempting to filter the useful information from the data. Therefore, instead of getting more data from the system, the system should give less information but of higher quality. The system described in this book is one which adapts the information to the personal need of a individual user, and which assumes responsibility for information filtering.

As part of a first approach to be presented, static information filtering is proposed which compares attributes of programs with given user profiles. An agent-based solution is based on examples of agent applications (e.g., [Markus 93], [Greif 94], or [Genesereth 94]).

The intelligent agent approach is applied to interactive television systems and introduces a novel approach of parsing and interpreting user interactions in the iTV end systems. The intelligent agent thereby provides easier and adaptive handling of daily tasks such as program selection, home shopping, and information retrieval. We introduce the term of the intelligent Media Agent (iMA) as the name for an intelligent agent in iTV systems. Decisions made by the agent are intelligent, personal, and adaptive. Agents are used to recognize the user needs, to translate this knowledge into personal user profiles, and to filter the spectrum of information being offered using this personal profile.

1.3.1 Motivation for TV Program Selection

Telemetrics research is primarily devoted to ascertaining when television programs are viewed by how many users. Motives for television program selection have been by analysing why the programs have been watched.

Gratification

[Gleich 95] gives an overview of the research results related to motives. Studies based on the "uses and gratification approach" (also see [Lin 93], [Schmitz 93], [Potts 94]) have shown that entertainment and information are the main reasons for watching television. Usually the results are based on psychological correlation of the program consumption to user requirements. However, it is very difficult to detect relationships between personal requirements and program consumption and vice versa. The question of defining a fundamental psychological model remains unresolved (see [Gleich 95]). A model defining the process of television viewing is given in [Lin 93]. He assumes that viewers are seeking gratification through television and that the fewer time a viewer switches programs, the more involved he is cognitively and emotionally, and the more his gratification need are being met. It has been determined (see [Lin 93]) that the behavior of user with high and well-defined expectations is more active, characterized by a more intensive program selection process. A deeper involvement in the program combined with a higher degree of post-processing of the information which results in greater satisfaction with respect to the degree of gratification obtained. No correlation has been observed between the frequency of television usage and user satisfaction.

In the studies mentioned above, it is pointed out that the reasons for watching television and programs are manifold. The following significant parameters can be defined: (1) favourite persons (e.g., actors, directors) are involved in the production of the program (see [Schmitz 93]); (2) specific interests of the viewer are present, e.g., the user likes the genre of the program; (3) the quality of the program corresponds to the gratification sought (see [Lin 93]); (4) the program is highlighted in show previews and in TV guides; (5) the programs entertainment value is high. [Opasch 94] has shown that television is also used to create an background atmosphere.

Costs

There is a trade-off between the gratification of television programs and their costs. The costs of viewing television are two-fold: money and time. The financial expense consists of the costs of buying the TV equipment, connecting to a TV distribution network provider, additional costs of subscribing to the TV program providers (e.g., ARD and ZDF), and costs for receiving additional Pay-TV channels. In interactive TV systems cost management will be more flexible. Given the elastic relationship between a program's cost and a users decision to view it, additional modes of cost management have been introduced in iTV and include:

Pay-per-pool:
A pool of channels (e.g., a broad mixture of special interest channels) is offered by the TV distribution network provider. The user pays for the set of channels and has access to these channels without extra payments.

Pay-per-channel:
Users subscribe to one TV channel. The subscription is valid over a certain period during which any and all programming can be viewed.

Pay-per-movie:
Similar to video rental stores, movies can be lent. The price paid is for a given time interval and is independent of the number of times the film is watched within this interval.

Pay-per-view:
The user pays for each view of the program.

Credits by advertisement:
Advertising revenue comprises the bulk of any TV channels budget. User benefit from this revenue in not paying for the programming on these channels.

These additional cost management strategies allow for more flexible and adaptive pricing. For example, the price of a movie could depend on the frequency and duration of advertising in the movie. Though the cost of a single program can vary, the cost structure constitutes a set of objective parameters which can be identified and utilized by a user for decisionmaking.

Incorporating new functionality at user interfaces is a subject of iTV systems research. Simple user interfaces are among the design goals of iTV navigation systems.

As a result, design specifications mandate that every user be able to control the iTV system. However, this approach does not take into account the individual requirements of experienced users (e.g., short-cuts for fast access, more complex operations at the user interface).

1.3.2   Program Search in Television Systems

In digital TV systems more than 50 television channels can be received by the users. In such a scenario existing strategies for channel selection can be expected to fail. Today, there are two major approaches to obtain TV program information:

(1) channel surfing; in which the user switches from channel to channel to get an overview of the programs which are currently running; and

(2) printed program guides; whereby the user studies printed program with descriptions and indices in order to get an a priori overview of the programming being offered. Assuming a TV service broadcasting 500 channels, it would take 83 minutes to surf over every channel once[2].

A printed program guide of the week would require about 350 pages a week. Both methods are unsuitable.

1.4   Need for Agent-based Systems

Agents, i.e. intelligent agents, are used to simplify user tasks. In the following, existing agent implementations are described in order to give examples of functionality applicable to interactive television systems. An overview of agent applications is given in [Brenner 97].

Search Agent / Information Broker

Search agents and information brokers are tools to search information domains and sub-domains. The functionality of search agents is similar to other well-known search engines. Examples of search agents are the intelligent Beluga meta-search engine [Fiedler 96] or the WWW search agent described in [Menzer 95].

---

2. It is assumed that the user needs 10 seconds per program.

Watchdog Agent

A watchdog agent is a typical example for the delegation and automation of tasks that currently require user involvement. The watchdog agent "watchcat" (see also [BTUC 97]) can be used to automatically observe HTML page modifications.

Navigation agents are responsible for filtering and personalizing information in applications where the amount of information given to users is very high. Workflow agents are used to automatize groupware applications (e.g., meeting agent).

Personal Electronic Program Guide

It is assumed that the individual and personal requirements of users must be regarded in the design and implementation of user interfaces in interactive television systems. It is necessary to adapt applications based on the specific requirements of individual users or user groups.

So far, electronic program guides have been identified as a key application in interactive television systems. The need to improve the quality of programs offered in the electronic program guide has already been identified. Therefore, an intelligent agent based electronic program guide — called the "personal electronic program guide" — is used as essential base to implement intelligent agents in an interactive television environment.

The concept and implementation of a personal electronic program guide introduces novel mechanisms for channel selection by using intelligent information filtering processes. An intelligent media agent is used for navigation and program collection in TV broadcast and video-on-demand scenarios.

The intelligent media agent is a software extension which resides in iTV application servers or end-systems and allows for the description of demands and wishes of users in terms of program categories, quality information, and statistical attributes. As a basic application it compiles a TV guide with a personal evaluation of programming.

1.5   Overview

This book describes the personal electronic program guide as a customized software application implemented within the intelligent media agent environment. Section 2 defines and classifies intelligent agents according to their specific role within the context of various iTV systems. Existing intelligent agent approaches, especially in the iTV environment, are compared.

Section 3 classifies and describes the user and system requirements related to the intelligent media agent. Two types of requirements are presented:

(1) requirements related to the multimedia run-time environment, and
(2) requirements related to the intelligent agent functionality.

Section 4 shows the design and implementation of agent run-time environments supporting multimedia user interfaces. Section 5 outlines the concept and architecture of the intelligent media agent, while in Section 6 user profiles and program descriptors are described.

Section 7 explains how program offers and user profiles are compared. Clustering is used as a basic algorithm to analyse the user interactions and to adapt the personal user profile based on the individual user's interaction. Various modes for initializing and updating user profiles are presented in Section 8. Because exceptions in the user behavior may occur a decision support module is introduced in Section 9.

In Section 10, those requirements which have been defined in Section 3 and which are met by the iMA are summarized. Section 11 outlines future extensions of the intelligent media agent.

2   Towards Personal Television Systems

This book is based on the fundamental assumption that users of information systems do not need more information, but information which better corresponds to their interests and needs, i.e. information of higher quality. This work describes a personal television application whose appearance is personally adapted to each individual user. In Section 2 fundamentals of personal iTV systems are given by discussing the definition of the term "intelligent agents" and by itemizing the specific tasks of personalization and the fundamentals of intelligent assistance. The functionality of an interactive, digital and personal EPG is described as an example of personalization, hereby addressing the specific problem of managing TV channels by using an intelligent agent. The following discussion is applied to a specific category of agents, namely software agents.

2.1   Software Agent Definition

The term *software agent* has often been used as a name for intelligent software modules. The work done in this field can be divided into two categories: the first concerning itself with the theoretical analysis of agent fundamentals and the second with the construction of agents acting within specific an focused user domains (see e.g. [Coen 94]).

There is no generally accepted definition of agents. To the agent theorists, agents are generally described as programs which imitate intelligent human behavior by doing things, making decisions, acting autonomously, and learning (see [Benyon 87],[Seel 89], [Shardlow 90], and [Chen 94] for more information on what agents are, how they work and interact). Laurel defines an agent as follows [Laurel 91]:

> "An agent can be defined as a character, enacted by the
> computer, who acts on behalf of the user in a virtual
> (computer-based) environment."

The term "intelligent agent" used in this book is defined as a autonomous program characterized by the properties in Figure 3 (also see [IBM 96]). Intelligent agents can be described by the three dimensions of agency, intelligence, and mobility. In the following, as a summary of related work in the area of intelligent agent theory, the attributes of an intelligent agent are given and explained (see [IBM 96], [Wilson 95]).

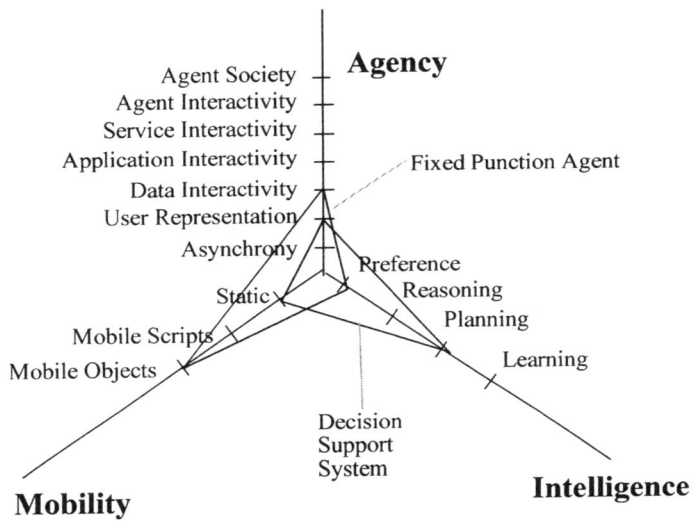

Figure 3: Intelligent Agent Properties

There is a threshold shown in Figure 3 which is used to distinguish intelligent agents from decision support systems and fixed function agents (e.g., system management agent).

2.1.1 Agency

The term agency reflects the interactivity of the intelligent agent with its environment. It is important whether tasks can be processed synchronously or asynchronously by the intelligent agent. A user representation is necessary to implement a model of the world in which an agent is acting. Data, application, and service interactivity means that these entities are able to communicate with other corresponding entities, e.g., other application or service modules.

Agent interactivity describes a distributed agent-based framework with inter-agent communication facilities. For example, intelligent agents can share rules and knowledge about their internal world model. The discussion of the interactivity aspects is closely related to the question of social behavior in societies of intelligent agents. In [Bouron 92] a social agents is defined "characterized by its capabilities to be integrated in a social context and above all by its capabilities to take an active part in the system organization". For example, in the agent interactivity category it is assumed that agents cooperatively exchange knowledge about the domain in which they are acting in.

2.1.2 Intelligence

The behavior of the agent should be intelligent. The question of defining the term "intelligence" is unresolved within the artificial intelligence research community. Often, the capability to survive is given as a measurement of intelligence. A system is as intelligent as its ability to maximize its property of survival in a given environment [Maturana 87].

Intelligence is an attribute describing the degree of reasoning and learning behavior. The intelligence of an agent depends on the abstraction level of instructions (=inputs) and the actions resulting from those instructions. The minimum scale of intelligence defines an agent whose behavior regards user preferences.

More sophisticated iMA's employ reasoning functions which are based on user models representing user preferences and which have the ability to plan and develop strategies.

For example, information filters can be used by agents to screen the information consumed by a user. Additionally, an agent can adapt this filter to include additional information sources which may be of interest to the individual user.

Higher on the scale agents are able to learn from their environments, i.e. are able to react to changes in the information base or in the user behavior by adapting its functionality to changes in the behavior and preferences of its user. Information discovery, the invention and exploitation of new concepts, and the ability to adapt the user model to these changes are further attributes of a learning agent. They are able to explore new information sources and derive new knowledge from their environment. The agent is able to develop exploration strategies and new representation models. Programming of the exploration functions can be expressed as follows [McFarland 92]:

> "To program an automatic agent means making him to do exactly what you as a designer want him to do; to program an autonomous agent means making him want to do what you want him to do [McFarland92]."

For example, structural changes in the information database (new hyperlinks in the worldwide web) must be recognized by an intelligent agent and the information filtering operations must be adapted to the new structure of that information.

2.1.3 Mobility

The scale of mobility describes the degree to which agents physically travel through the network and perform action autonomously. Static agents reside on a fixed set of computers (minimum = 1) with a fixed set of predefined functions for each machine.

Mobile scripts are composites of agent instructions which are sent to and executed on another host computer. The process of sending and executing mobile scripts is stateless.

Mobile objects contain agents or parts thereof which allow the remote execution of agents with states. Mobile agents are transported from computer to computer. One such example is the mobile shopping agent which sells wares in a virtual shopping marketplace (see [MMS 97]). Therefore, an intelligent agent's autonomy is closely related to the mobility of its parts.

Autonomy

The term "autonomy" is derived from "autos" (gr.: self) and "nomos" (gr.: policy, rule): the definition and necessity of autonomous instead of automatic activities has been motivated by McFarland [McFarland 92]:

> "An Agent whose behavior is entirely controlled by an outside agent has no "will of its own", or self government. In everyday usage, such agents are called automata, their actions being "involuntary"."

and

> "Autonomous Agent are self controlling as opposed to being under the control of an outside agent. To be self-controlling, the Agent must have relevant self-knowledge and motivation, since these are the prerequisites of a controller. In other words, an autonomous agent must know what to do to exercise control, and must want to exercise control in one way and not in another way."

An example of an autonomous agent is described in [BTUC 97]. Here, the autonomous agent helps to reduce interactions between users and the agent system by extending the agent activity in the network.

2.2 Related Work

Basic intelligent agent technologies can be subdivided into four groups:
1. Neuronal Networks (see [Arbib 87], [Cruse 93], [Kosko 92], [Pfeiffer 92])
2. Algorithmic Approach (see [Brooks 86], [Brooks 90])
3. Circuit-based Approach (see [Kaelbling 87])
4. Dynamic Approach (see [Steels 94])

An detailed overview and description of the above approaches are given in [Vazirian 95]. Further classification criteria and examples are given in [Brenner 97]. Additionally, reasoning and learning approaches taken from the AI research have often been used as the basis of an intelligent agent implementation.

2.3 Practical Issues and Examples

The concept of intelligent agents can be merged with existing approaches of multimedia systems. By combining intelligent agents with multimedia systems, agent applications which intelligently compose multimedia information can be created. An agent which helps to manage media processing is called an intelligent media agent.

In an interactive television environment an intelligent media agent is responsible for the following tasks: (1) the adoption and/or adaptation of user interfaces and dialogs, (2) the support of tele-shopping and other interactive services, and (3) automation of information retrieval, and (4) personal movie indexing and generation. The different tasks are handled by autonomous parts of the iMA, the so-called subagents, or softbots. The following describes these tasks in greater detail.

2.3.1 Adoption of User Interfaces/Dialogs

In future iTV systems, support for intelligent user interfaces is very important. As in traditional television systems, users come from all social strata, have completely different backgrounds, and have no particular technical training. Because there are many users who do not know the first thing about computers or VCR programming, the design of user interfaces for iTV systems must be different from today's interfaces. Based on common user interface design principles (e.g., in [Albers 63], [Laurel 91], [Carrol 91]) interfaces must be adjustable to the properties of users. Knowledgable people are provided with an extended set of control functions, whereas beginners will be given simple interfaces. That part of the iMA responsible for user interface design is called the user interface softbot.

The control of user interfaces is an important sphere of work in artificial intelligence research. Fundamentals can be found in [Kra94] and [Rouse 87]. The variety of functions ranges from intelligent help functions (e.g., [OMalley 86], [Erlandsen 87], and [Rosenberg 81]) which are activated using timed triggers, to high-end functions with knowledge of the properties, and behavior of users (e.g., [Markus 93], [Benyon 87]). In iTV systems, the iMA records actions and waiting times of users. Because the behavior of users reflects their personal properties, this information can be used to build up an internal knowledge base about the user. By applying common rules about user interface design and page layout, the user interface can be automatically designed depending on personal requirements.

For example, elements of a dialog could be changed if the intelligent agent notices that a button in the upper left corner of the screen is never used. From a user interface rule base, it can be seen that this button is for novice users (e.g., the tutorial help button). In this case, the iMA draws two conclusions, namely that the dialog element can be hidden because it has never been used, and, that it is very unlikely that it will be ever used in the future. Additionally, the iMA knows that this user is likely to be advanced, because he has never needed assistance.

However, rules should be applied very carefully (for example, the user may have hard-copy manuals). Developers and designers of iTV presentations can define fixed elements for the basic structure and variable elements which are filled with personal data by the iMA.

2.3.2 Shopping and Other Interactive Services Support

Interactive home shopping, banking, and a mixture of various on-line services are one of the driving forces for commercial interests in iTV systems. The iMA helps to navigate through virtual malls and builds such virtual malls according to the wishes and budget of the user. It helps to seek special offers or to find the best price for an article by comparing catalogs of service providers. The iMA can suggest matching taking into consideration both past shopping habits and the current prices. An agent can help to reduce costs and adapt the shopping behavior to the particular budget of the household. The subagent which optimizes of shopping and service access is called the shopping softbot.

The iMA also supports distance learning and tutoring by adjusting the speed and degree of difficulty ([Maes 93a], [Clancey 87]). The iMA should also be extendable by specific subagents, for example, assistants for the stock exchange which employ artificial intelligence methods to chart and analyze data fundamental as well as predict stock prices (e.g., [Hutchinson 94]) via the iTV end-station.

2.3.3 Software Agents for Information Retrieval

Interactive TV systems can also serve as the front-end for interactive information retrieval in global networks. Because there is a large number of ways to get the right information, and these ways are rarely structured, beginners need much more time than an expert user to accomplish the same task. Users usually need a learning process which gives them experience and knowledge in effective search strategies (meta-knowledge). Successful searches have as a consequence that a certain search strategy will be recognized as having been effective. Similar to lists of indices in the Internet, meta knowledge about indices (2nd level meta-knowledge) helps to find specific information faster than direct searches in the 1st level meta-knowledge base. Various approaches to the information retrieval problem can be found in the literature (for example, [Alberico 90], [Kehoe 85], [Lochte 93]). In most cases, the core is a decision support system which contains a knowledge base with search algorithms constructed by an expert in the specific field. A general model for an information retrieval software agent is given in [Vorhees 94]. The information retrieval part of the iMA is the retrieval softbot.

2.3.4 Personal EPG and Movie Generation

Types of Movies

The portion of the iMA which is responsible for all tasks dealing with movies is called the movie softbot. There are three types of movies in interactive TV systems: (1) linear movies, (2) interactive movies, and (3) variable movies.

Linear movies describe the most prevalent form of movie shown today. Linear movies are best-suited for broadcast TV, where the users cannot influence the broadcast. There are two approaches to get an overview of TV programming. A very popular method is channel zapping, where users switch from channel to channel to see the current programs. The other alternative is to browse TV guides which contain indices to the programs on the TV channels. An example in Section 1 has shown that it is very difficult to overview the program when the number of channels exceeds a certain value. Then the search time is very high compared to the search times today.

One solution to the problem is the system proposed and developed in this book: Information management using the iMA helps to filter out the useful information and to restrict and minimize the information to an amount which can be processed by the users. A major application is a personal EPG containing only programs which are of interest for the user. The iMA allows for the description of demands and wishes by users, converts them into personal profiles, and applies these to the EPG. In addition to the simple off-line mode where the users specify their profiles, the agent applies mechanisms for automatic profile generation by watching the operations of the users (e.g., channel switching, duration of watching specific programs). The iMA uses this feedback to check its offerings. For example, the agent should automatically correct the profile when a user mostly watches football games but the user did not specify this interest in the personal profile. Thereby, the iMA can optimize the TV guide. Another aspect of optimization comes with the costs of certain channels or movies. Users want to be aware of the costs of their TV consumption but do not want to be constantly asked "The next program will cost 10 cents. Is this OK?". The iMA can manage a monthly budget for the user. The user specifies what maximum amount of money he intends to consume. Based on the personal interests and the restricted payment, the iMA browses through the channels and assembles an optimized TV guide.

Another task of the movie softbot is the optimization of videos in the Near-Video-on-Demand mode. Near-video-on-demand is a specific broadcast mode for linear movies with slotted access. The starting time of one movie on different channels is shifted by a given number of minutes. It is up to the user to choose the starting time in the resolution of time slots. This enables users to change positions in the movie, and to view parts of the movie again. Here, the intelligent media agent is used to optimize the starting times of programs. For example, if the agent finds out that the broadcast times of the users most favourite movies X and Y overlap that day, it tries to take movie X as early as possible and movie Y as late as possible. The agent determines whether any possibility exists for the user to see both movies.

Interactive movies are expected to appear in the first generation of interactive television systems. At pre-defined times branches appear in the story of the movie, where users can interactively decide what sequence should follow. For example, it could be up to the users to decide whether the hero must die (and the movie is tragic), or the scoundrel (and there is a happy end). According to the structure of the TV distribution network and systems, either there is a competition of personal opinions among viewers (multicast and broadcast networks), or the single viewer can choose the scenes (unicast networks). However, the user can not be expected to make a decision every minute. Therefore, the story should branch after long intervals, and these branches have to be defined carefully. This can be done by the iMA. The iMA reduces the number of branches by making the unimportant decisions on behalf of the user.

Live transmissions are another opportunity for interactive TV. In this mode, users can be their own directors of the live event. They choose the camera positions, initiate interviews, or ask for instant replays. Similar to the interactive movies, users should not be compelled to write a directing guide before he can see the sports event. He can either choose to follow the standard direction of the TV channel, or delegate the minor decisions to the iMA, making all important decisions himself.

Variable movies or cybervision movies exploit the advantages of virtual reality [Sack 94]. This kind of movie is not expected in the next generation of iTV systems. However, in the future it is expected that iTV systems, or at least home computers, will run virtual reality games or movies. It is a dream of human beings to take part in adventures, to play roles in movies, to influence the action, and to have the chance to become the hero. In this mode of viewing, personalization is most important. The scenes are dynamically created in reaction to the behavior of the actor.

2.4   Intelligent Agent Research Applied to Television

Based on the careful review of previous research in the areas of television systems and intelligent agents, research applied to the combination of television systems and intelligent agents will be explored in this section.

In digital television systems a navigation tool is usually responsible for guiding the television user and displaying program information. Examples of such navigational tools are the navigator in the Nuremberg (Germany) iTV trial (see [MMS 96a]) or the navigator for the digital German broadcasting bouquet DF1 (see [DF1 97]). Starsight and Prevue Interactive are electronic program guides (EPG's) which are offered in US television cable networks. Though easy navigation through television programming is a general requirement, personal EPG's have not yet been offered so far. There is no interactive television system known to the author which implements intelligent filtering and clustering operations including the adaptation of profiles by using a reasoning and learning approach.

Examples of systems which contain recommendation algorithms are given in Table 2 (for more details see [Fisk 96]).

Table 2: Related Recommendation Systems

System	Domain	Comments
Movie Select	Movies	see below
Movie Recommendation Engine [CMU 96]	Movies	No information on how the systems works is publicly available.
Firefly [Agents 96]	Movies, Music	This system was developed by the Autonomous Agent Group at MIT Media Labs. It uses social information filtering.
Movie Recommendation System	Movies	see below
GroupLens [Resnick 94] and [UMN 96]	Usenet News	Developed at MIT Media Labs, based on social information filtering.
NewT [Maes 93b]	Usenet News	Developed at MIT Media Labs, based on genetic algorithms, with further non-genetic learning taking place between generations
Siemens Personal Newspaper	News	see below

In the following text, some approaches for personalized information filtering are considered in more detail and compared with the concepts of a personal electronic program guide in interactive television systems.

Gleichzeitig bestelle ich zur Lieferung über meine Buchhandlung:

Expl.	Autor und Titel	Preis

Weitere Informationen finden Sie im Internet:
www.vieweg.de

Verlag Vieweg –
Einer der ältesten Verlage der Welt.
Gegründet 1786 in Berlin.
Partner von über 30 Nobelpreisträgern.

ALBERT EINSTEIN
14.3.1879 – 18.4.1955
NOBELPREIS FÜR PHYSIK 1921

Antwort

Friedr. Vieweg & Sohn
Verlagsgesellschaft mbH
Leser-Service
Abraham-Lincoln-Str. 46

65189 Wiesbaden

Ich interessiere mich für die Themen:

- ❏ Mathematik (H5)
- ❏ Informatik (H5)
- ❏ Wirtschaftsinformatik (H55)
- ❏ Computerliteratur/Software (H55)
- ❏ Physik (H5)
- ❏ Chemie (H5)
- ❏ Architektur (H9)
- ❏ Bauingenieurwesen (H6)
- ❏ Techn. Mechanik (Bauwesen) (H6)
- ❏ Bauphysik (H6)
- ❏ Werkstoffwissenschaften (H6)
- ❏ Techn. Mechanik (Ingenieurwesen) (H6)
- ❏ Technische Thermodynamik (H6)
- ❏ Maschinenbau (H6)
- ❏ Elektrotechnik (H6)
- ❏ Kfz-Technik (H6)
- ❏ Umwelt-Techniken (H5)

Ich interessiere mich für folgende Produkte:

- ❏ Bücher
- ❏ Zeitschriften
- ❏ Computerunterstützte Lernprogramme/PC-Trainer
- ❏ CD-ROM/Anwender-Software

Ich wurde auf dieses Buch aufmerksam durch:

- ❏ Empfehlung des Buchhändlers
- ❏ Empfehlung Kollegen, Bekannte
- ❏ Buchbesprechung/Rezension
- ❏ Anzeige/Beilage
- ❏ Werbebrief

Ich bin:

- ❏ Dozent/in
- ❏ Lehrer/in
- ❏ Bibliothekar/in
- ❏ Sonst. _____
- ❏ Student/in
- ❏ Praktiker/in
- ❏ Schüler/in

an der:

- ❏ Uni/TH
- ❏ FH/HTL
- ❏ Fachsch. Technik
- ❏ Berufsschule
- ❏ Gymnasium
- ❏ Bibliothek
- ❏ Sonst. _____

Mein Spezialgebiet: _____

Bitte in Druckschrift ausfüllen. Danke!

Hochschule/Schule/Firma _____   Institut/Lehrstuhl/Abteilung _____

Vorname _____   Name/Titel _____

Straße/Nr. _____   PLZ/Ort _____

Telefon _____   Fax _____

Branche _____   Geburtsjahr _____

Funktion im Unternehmen _____   Anzahl der Mitarbeiter im Unternehmen _____

Wir speichern Ihre Adresse, Ihr Interessensgebiet unter Beachtung des Datenschutzgesetzes.

MBVM

2.4.1 Movie Select

Movie Select [Medior 95] is a database containing 44,000 movies stored on a CD-ROM. As stated in the manuals the search algorithm is claimed to incorporate search functions, artificial intelligence, and fuzzy logic. Further details about the internal algorithms have not been published. There is no online version, and the database does not contain the TV program lists. Reasoning and learning algorithms are not applied. There is no internal representation of the user preferences.

2.4.2 Movie Recommendation System (MORSE)

MORSE (see [BT 97] and [Fisk 96]) is an Internet service developed by British Telecommunication and provides recommendations for movies. Personal ratings of selected movies are used as input into the rating system. This input is used in two different manners as follows: First, personal user ratings are used to search for interesting movies. Figure 4 shows an example of individual movie recommendations

Second, the system accumulates an overall statistic for each movie and displays it on a chart listing the most popular movies. Details of the method for comparing the personal ratings with average ratings of the movies in the movie database have not been published. The result of that comparison is a list of top movie recommendations.

It is not clear whether the comparison function utilizes a reasoning approach. The calculation of preferred movies is made using preferences which have been put into the system by a sample selection of movies which have been rated by other users. Currently, there is no integration of the internet-based MORSE approach and the interactive television system. There is no mechanism to take into account the user's movie-watching behaviour in order to check and update the user profile.

2.4.3 Siemens Personal Newspaper

TheSiemens Personal Newspaper system is a prototype system which creates a personal newsletter utilizing personal preferences of an individual user. The system is based on 8 pre-defined profiles which can be selected by the user [Siemens 95]. Static filtering operations employing fixed attributes are used to calculate the pre-defined profiles.

	score	± error	count	mean	film
					Predictions for hw@mms-dresden.telekom.de:
weakly recommended	10.00	± 4.24	38	7.76	The Maltese Falcon (1941)
recommended	10.00	± 3.14	36	8.14	Cinema Paradiso (1988)
weakly recommended	10.00	± 4.56	43	8.58	The Shawshank Redemption (1994)
weakly recommended	10.00	± 4.29	29	8.28	North by Northwest (1959)
	10.00	± 6.02	82	7.04	Picnic at Hanging Rock (1975)
recommended	10.00	± 3.05	33	7.09	Howards End (1992)
	10.00	± 7.30	23	7.35	If... (1968)
recommended	10.00	± 3.89	62	8.11	Vertigo (1958)
	10.00	± 5.23	22	8.09	Delicatessen (1991)
	10.00	± 6.87	206	8.42	The Wrong Trousers (1993)
recommended	10.00	± 3.22	187	7.50	Taxi Driver (1976)
weakly recommended	10.00	± 4.28	378	7.79	One Flew Over the Cuckoo's Nest (1975)
weakly recommended	10.00	± 4.86	28	7.39	The 39 Steps (1935)
recommended	10.00	± 3.39	167	7.96	Rear Window (1954)
weakly recommended	10.00	± 4.91	94	7.79	The Third Man (1949)
recommended	10.00	± 3.00	36	7.33	Raise the Red Lantern (1991)
weakly recommended	10.00	± 4.49	253	8.31	Schindler's List (1993)
	10.00	± 6.87	48	7.38	The Sting (1973)
	10.00	± 6.51	23	7.30	The Double Life of Veronique (1991)
	10.00	± 6.10	40	7.72	Three Colors: White (1994)
recommended	9.96	± 3.82	36	7.89	Ed Wood (1994)

Figure 4: MORSE Recommendations Screen

2.4.4   Evaluation

Static information filtering is used as method for personalization. The user must specify his information profile. The process of filtering is based on fixed methods. Adaptation can only be achieved by changing the filter by hand. There can be no adaptation of filters or automatic recognition of user interests. In interactive television environments this approach would fail because it does not provide automatic installation and update modes which are necessary to support users who have traditional TV interfaces.

Social information filtering is a popular method used in recommendation systems. The general method is as follows [Fisk 96]: The user rates a media (e.g., movies, music), and those ratings are compared with the ratings of other users. If there are similarities with the tastes of others, the system recommends those media of the others which were highly rated by others of similar taste. Social information filtering is a frequently used method in interactive systems. So far, the algorithms used in social recommendation systems seem to be applicable to the problem of personalized program guidance. However, it requires interaction between the user and the social recommendation system.

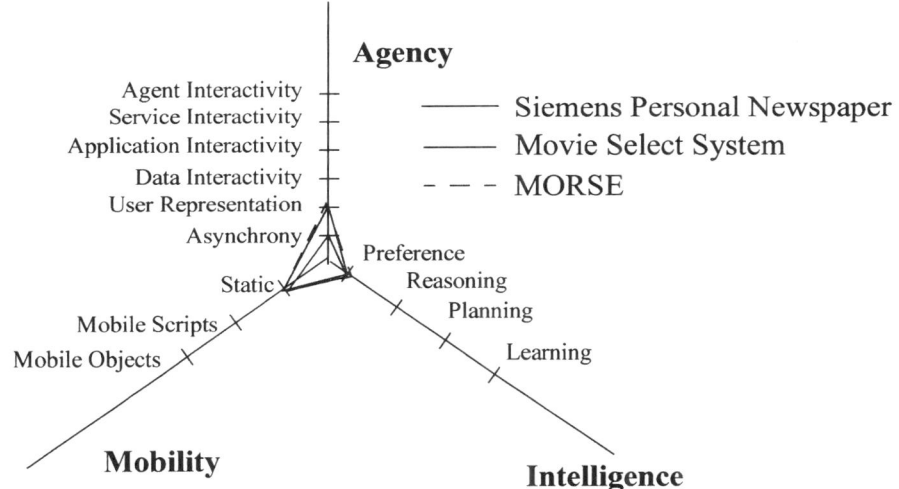

Figure 5: Overview of Existing Approaches

At a very minimum, personal preferences related to the application domain have to be specified by the user and given to the social recommendation system. This requires an interface which is more complex than the television interface of today. Therefore, there is a conflict between the essential usability goal of the personal EPG application (i.e., to use the current TV interface) and the functionality provided by social recommendation systems. This type of filtering and specification is recommended to those users who are able to install and manage their personal ratings. Social information filtering is unfeasable for those users who insists on using the same interface. Most social recommendation systems are based on user input which not necessarily reflecting the user's real watching behaviour. However, this behaviour is not taken into account when recommending movies to an individual user. By contrast, the personal EPG application should be a system which does consider when and how long programs are watched as indices of personal preferences. This assumption is the result of media research as already presented in Sections 1 and 2.

The spider diagram in Figure 5 depicts the properties of the existing approaches. MORSE is the most advanced system known to the author. Most of the approaches in the area of interactive television do not contain reasoning or learning functions which are necessary to implement an intelligent functionality.

Systems which focus on the selection and presentation of individual information in the area news and electronic shopping systems is presented in [Brenner 97].

3 Requirements

Requirements to the intelligent media agent can be categorized into those requirements which are related to the multimedia run-time environment, and those requirements related to the core intelligent agent functionality.

3.1 Requirements to Multimedia Environments

It can not be assumed that each user is familiar with traditional computer-style interfaces. Therefore, the intelligent media agent should employ a standard multimedia user interface. This interface should also be adaptable to the personal requirements of the user. The following characterizes the run-time environment requirements for the intelligent media agent.

3.1.1 Customizing

Users possess different technical backgrounds. It is important to have the ability to be compatible with the technical knowledge of a given user.

R1 The multimedia user interface can be customized.

3.1.2 Portability

The multimedia run-time environment consists of heterogeneous operating systems and hardware platforms. One major insight gained by the iTV field trials is that standards only at the monomedia level of information (e.g., MPEG-2, JPEG) are insufficient to guarantee application portability in iTV systems. Monomedia standardization also does not address the interplay between multimedia and hypermedia information (i.e., the look and feel of the application). In order to deliver a personal EPG to a variety of heterogeneous operating system platforms, standardized multimedia application run-time interfaces are required.

R2 An interface providing portable applications is offered by the agent run-time environment.

3.1.3 Modifiability

When presenting a multimedia application it is required that this multimedia application be able to be dynamically changed. Due to the dynamic properties of the personal electronic program guide, modification and adaptation of the multimedia application or parts of the application should be provided.

R3 The multimedia application is modifiable and adaptable.

3.1.4   Multiple Application Support

Due to the resource constraints (e.g., low power or memory) of the set-top boxes, it is desirable that an intelligent media agent be concurrently used by a set of agent applications (for the types and tasks of intelligent agent application see Section 1.2).

R4   The intelligent media agent kernel provides multiple agent applications.

3.1.5   Distributed vs. Local Processing

Depending on the network infrastructure and the processing capacity, local and/or distributed processing of agent functionality is required.

R5   The agent's run-time environment supports distributed and local processing.

3.2   Requirements to Program Provision

In addition to the multimedia run-time environment, the intelligent media agent kernel is responsible for implementing functions which let the personal electronic program guide appear intelligent. The definition and specification of programs and their provision are all important functions and reflected by defining related requirements.

3.2.1   Program Providers

An electronic program guide is an application providing access to a database containing the programming information for one or more TV networks. While the majority of this information is constant, a small percentage is dynamically updated with last-minute changes as, for example, in coverage of special live events. Therefore,

R6   The intelligent media agent application regards multiple TV channels and TV program information.

R7   Dynamic changes in the TV program is provided (i.e., "last-minute changes").

3.2.2   Program Description

Television programs can be characterized by a set of attributes. For example, traditional television guides provide information about the actors and the script writers together with textual description of the program.

R8   The intelligent media agent provides a description of the television program.

R9   The program description contains statistical information.

R10   The program description specifies the genre or category of the program.

R11   The program description specifies the quality of the program.

3.2.3   Fuzzy Specification

In the specification process, a user may have requirements whose attributes are not precisely determined or can not be completely or absolutely expressed by users. These attributes can be set by using fuzzy specifications. An introduction into fuzzy logic is given in [Bezdek 93] and [CMU 97]. Fuzzy sets were introduced by Zadeh (see [Zadeh 65]) as an approach to express fuzzy parameters and to manage vagueness.

The genre of a program, its quality and era are examples of fuzzy input parameters. Assuming a user is interested in movies which were produced in the rock and roll era, this parameter must be defined based on a fuzzy description (see Figure 12). It cannot definitely be said when the rock and roll era begun or ended.

R12   Fuzzy attribute specifications are supported.

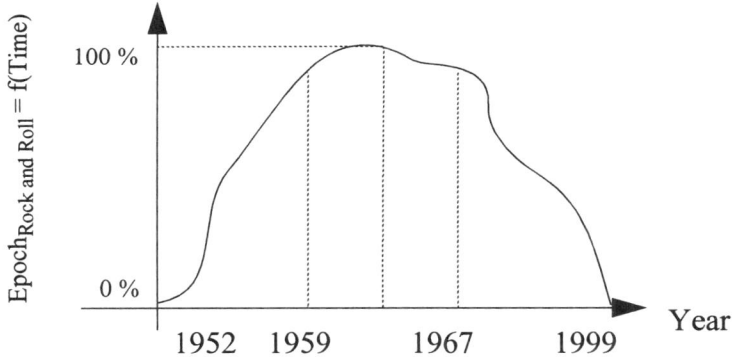

Figure 6: The Rock and Roll Epoch as Fuzzy Attribute

3.3   Requirements to User-Specific Functions

The requirements related to user-specific functionality can be subdivided into those addressing the specification of user profiles and those addressing their processing.

3.3.1   Specification of User Demands

Statistics have shown that there are various motives for selecting and watching television programs. For example, genre can be a criterion for program selection, one instance of this being the genre preferences expressed by children as described in [Krüger 96b]. Program costs can also influence the user selection of the program (e.g., see pay-TV [Zimmer 96] and pay-radio [Breunig 96] forecasts). The intelligent agent should therefore be based on a model implementing the various user requirements related to the program selection.

R13   The intelligent media agent define a user input model which represents the user interests and determines the user demands related to television programming.

Program preferences and the amount of time spent watching television varies on the day of the week. For example, on average, a German household watched 84 minutes of television Monday through Friday, and 114 minutes Saturday and Sunday. Since user interests vary directly with the time of the day, day of the week, and holidays (e.g., see [Darschin 96], [Feier 96]), it is required that:

R14   Time of the day, day of week and holidays be regarded when registering user demands.

Regarding Wishes and Demands of Users

In order to generate a personal TV program schedule, it is required that users define their preferences. Because these preferences are usually expressed in terms of the current offering of the television programs (see [Darschin 96]), the specification of the user profiles should closely be oriented to the descriptors of the TV program which have been described in requirements R8 to R11.

R15   A specification of user interests based on program attributes is provided by the intelligent agent.

It is assumed that the strength of a given program preference can vary depending on both the user and the program (see [Schmitz 93], [Opasch 94]). The specification of favourite programs can be mandatory or optional. This variability should be provided for in the specification of user profiles and the calculation of the personal electronic program guide.

R16   A specification of the degree of program preference (e.g, wishes and demands) is required.

Profile Specification Modes

Television services are used by almost every household (98.5% in 1994, see [Media 94b]), varying degrees of technical competence being represented by this group.

It can not be assumed that users are familiar with traditional computer interfaces (e.g., see [Weiler 97]), and a significant group of users will require that the interface to the intelligent agent is as easy, or almost as easy, as today. New television navigation systems would not be accepted by these users. Therefore, a broad set of various profile specification modes should be provided which can be used to adapt the method of profile specification to the individual technical knowledge of the users.

The modes are proposed as follows:

R16.1   *A user can explicitly specify his program interests.*

R16.2   *A user can specify select individual properties (e.g., age, profession) from which the iMA should be able to derive basic user interests.*

R16.3   *A user can specify a representative list of appealing programs from which the iMA can induce user interests.*

R16.4   *The iMA can automatically infer user interests by observing user interactions for a certain time.*

Specific attention should be given to requirement R16.4 because this mode can be expected to be typically used by all users who do not want to use a more complex interface than today's.

R16.5   *The specification procedure is adaptive.*

R16.6   *The results of the specification is made available immediately.*

R16.7   *The results of the specification phase reflects the interests of the user.*

3.3.2   Processing and Output

Portable Applications

Intelligent agent applications should be portable. Therefore, the portable application interface required in R2 must be supported.

R17   *The output (=application) of the intelligent media agent is based on a portable multimedia application format as required in R2.*

Multiple User Views

There are different kinds of information access. For example, suppose that a user is interested in all programs related to computers. In this case the agent is required to allow the specification of this requirement (see also R13) and to support a specific view displaying only computer related programs. Another view mode is required if a user has the time to watch television,

and if he is interested in a list of topics which are related to his set of interests.

R18   *The intelligent media agent supports the processing and application of multiple views to the television program data.*

User Groups

User groups can be interested in finding programs which are of interest to all group members.

R19   User groups are supported.

Management

Due to the broad spectrum of people watching television, it is required that a spectrum of user interfaces be provided. This implies that in a basic mode for novice users, no administration of the intelligent media agents is expected. Advanced users may access the administration data of their intelligent media agent.

R20   There are different modes of administration ranging from no administrative capabilities to full administrative capabilities.

Update of User Profiles

The update of user profiles is necessary if the user preferences change. It is very likely that the user interests will change over the time. Therefore, the intelligent media agent should be flexible when specific user interests do not exist or new interests arise.

R21   The iMA provides facilities to update user interests.

Similar to the functions which are generally required for the management of user interests, (also see requirement R16) the following update modes are required:

R21.1   A user is able to explicitly specify new programs of interest and delete old programs. This mode is specifically required by expert users who are familiar with the specifics of the update capabilities.

R21.2   A user can change selected individual user-related properties which are used to derive basic user interests.

R21.3   A user can update his representative list of programs.

R21.4   A user can specify that the iMA automatically update his interests based on the iMA's observations of his interaction.

The automatic update mode is expected to be the preferred mode of all users who want to retain their same user interface. In this case the intelligent media agent is transparent to the users until the output is generated. In this mode the recognition of user interests is automatically made by the agent. Normally, a change in user preferences should cause the system to modify the user profile.

The quality of that process is detemined by the degree to which the updated user profile accurately reflects the real change in user preferences. In the best-case scenario the two are identical.

R21.5   *In the automatic update mode, changes in user interests are equivalently represented in the user profile of the system.*

For example, minor changes in user interest should not result in major changes to the user profile managed by the system.

Exploration

In general, the interests of the user may change over time. The exploration of new interests should not be prevented by the intelligent media agent. It is therefore important that the intelligent media agent account for the ability to explore new television programs and channels.

R22   *The programming offer to a user should not be comprised solely of his favourite programs. It should permit new interests to be pursued.*

Processing

In order to reflect the completeness in the processing user profiles,

R23   *the iMA regards each attribute in the user profile when calculating the personal electronic program guide.*

Result

With respect to the iMA's task of filtering out programs of interest to the individual user, the following requirements are need to be met:

R24   *The result is stable, i.e., similar input values to the processing results in similar outputs.*

R25   *The result reflects the personal preferences as precisely as possible.*

Exception Handling

When initializing and updating the user profile automatically, the intelligent media agent is responsible for updating the user profiles on behalf of the user. For example, this can be done by using the remote control. An intelligent agent should provide mechanisms which detect exceptions when interpreting the user feedback.

One example of exception handling is a situation where a child visits his grandfather. The television programming watched by the child differs from that of the grandfather. Normally, in an automatic agent mode, the personal electronic program guide would contain the interests of both the child and the grandfather, even after the child has left the household. By using an exception handling mechanism, however, the iMA can recognize rapid changes in the user profiles and detect the end of such a visit, thereby recovering the previous user profile of the grandfather.

R26   *When interpreting user feedback the intelligent agent takes care of exceptions.*

Mobile Personal Profiles

27.1% of all German households are equipped with more than one television set [Media 94b]. As a requirement for the intelligent media agent, users should be able to transfer their personal user profiles to other TV sets.

R27   User profiles should be mobile.

Independence

Because a user has entrusted the iMA to act in an unbiased manner on his behalf, it is important that the compilation of personal TV guides not be influenced by network programming providers.

R28   The iMA regards different TV channel providers on a fair basis.

R29   The calculation algorithm of the iMA is unbiased.

3.3.3   Privacy and Security Issues

It is essential that users have confidence that no unauthorized access will be granted to the sensitive private data stored in their user and usage profiles. User and usage information should be confidentially stored. No access to the user profile and usage data information should be allowed without prior permission of the user.

R30   The private user and the usage data is protected from unauthorized access and the privacy of user profiles must be guaranteed.

3.4   Conclusion

This section has defined the set of requirements which must be regarded by the intelligent media agent. In the subsequent sections it is shown how these requirements can be either completely or at least substantially met.

4   Agent Run-Time Environment

The agent run-time environment is comprised of all software entities interfacing to the intelligent media agent. This environment provides substantial interfaces for multimedia presentations, dynamic presentations, or for a secure environment insuring privacy.[3]

It will now be shown how the suggested implementation of the agent run-time environment is suited to fulfil the requirements R1, R2, R4, and R30. Two levels of abstractions related to the run-time environment abstraction are defined: (1) Systems standards comprise the definition of all services and interfaces from the content source via content provider, service provider and service user. A standardized environment results defining all functional entities required in an interactive television environment. (2) Multimedia standards define control and management interfaces related to the presentation, representation and manipulation of the multimedia data at the user interface.

In this section, important system and multimedia standards are given, and components of the run-time environment are presented.

4.1   System Standards

International standardization committees have standard system architectures and run-time for interactive television systems defined, namely, (1) Digital Audio-Visual Council (DAVIC), and (2) Digital Video Broadcasting (DVB).

4.1.1   DAVIC

DAVIC ([Chiari 95], [DAVIC 95d]) is both the name of a consortium and of a system standard for broadcasting and interactive television systems. DAVIC includes multimedia standards which are used for application development in DAVIC compliance systems. The DAVIC system is comprised of the following components of an interactive digital interactive television system: clients, an application server, a video server, system management, and gateways. In Figure 7 the architecture of an System, a Service Provider System, Service Management components, and Servicew Consumer Systems.

---

3. Parts of this section were published as [Cossmann 96], co-authored by the author of this book.

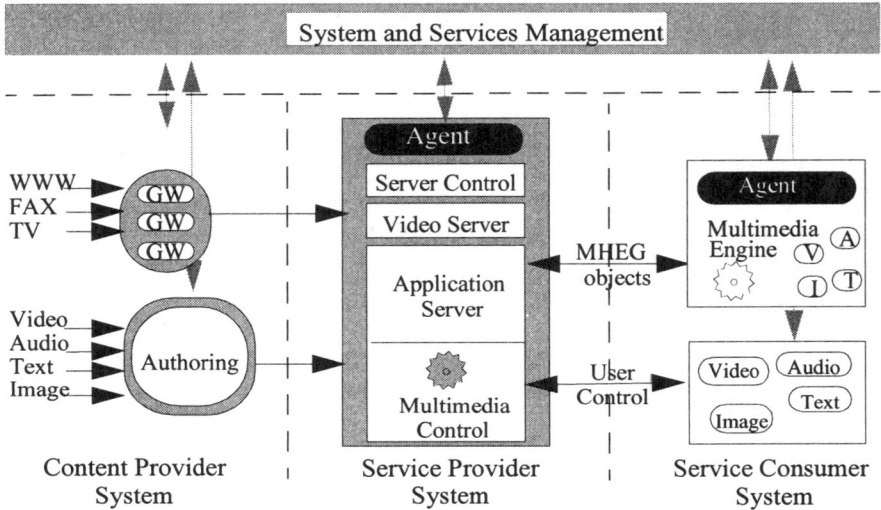

Figure 7: Architecture of a DAVIC Compliant System

In the Content Provider System, digital monomedia data formats (e.g, MPEG-2 video as well as audio, text, and images) can be processed. A multimedia application is built by combining monomedia contents, e.g., by adding temporal and spatial synchronization information. Presentation, representation and manipulation information is added to the monomedia components. The resulting multimedia applications and monomedia contents are sent to the Service Provider System. It is comprised of video servers for the playback of audio and video contents, of application servers which run the user application's back end, and of control processes for server management. The multimedia presentation processing takes place in the Service Consumer Systems. In the context of agent-based applications, the intelligent agent can reside either on the Service Provider System or in the Service Consumer System.

4.1.2 DVB

The DVB project standardizes digital television broadcast systems. Details are given in [Reimers 95a] and [Reimers 95b]. The DVB standard is focused on the transport of digital video over satellite, cable and terrestrial networks.

Additionally, standards have been defined in the subscriber management/access control and services information domains.

4.1.3   Multimedia Standards in DAVIC and DVB Compared

The DAVIC consortium chose the *ISO/IEC Multimedia and Hypermedia Experts Group (MHEG)* standard as the portable multimedia application format. In the DBV system standard, there is no standard defined for interactive services and multimedia application formats. The following sections contain a brief description of the components required for developing an electronic program guide based on the MHEG and HTML/JAVA standards. Thereby, it is shown how requirements R2, R4, and R30 are fulfilled by the personal electronic program guide implementation in a DAVIC standard compliant environment.

4.2   Multimedia Standards

In the context of the intelligent media agent the standardized multimedia environment is responsible for providing data, as well as control and management services related to the presentation, representation and manipulation of multimedia user interfaces. A standard multimedia format is necessary in order to provide a scalable solution for assembling, encoding, transmitting, decoding and finally presenting the electronic program guide. In the following, two standard multimedia formats, MHEG and HTML/JAVA are introduced as a potential base for the multimedia environment used in the intelligent media agent.

4.2.1   Multimedia and Hypermedia Experts Group Standard

MHEG is the name of the "Multimedia and Hypermedia information coding Experts Group". This group is organized as working group 12 of the ISO/IEC Joint Technical Committee 1/ Sub-Committee 29. The standardization within MHEG is split into several parts. Part 1 of the MHEG standard (MHEG-1) is the description of the "Coded Representation of Multimedia and Hypermedia Information Objects (MHEG)" in the Abstract Syntax Notation 1 (ASN.1). Due to the restricted resources of a set-top box, in part 5 of the MHEG standardization (MHEG-5) the subset of MHEG-1 objects for digital and interactive television applications has been defined. There are liaisons between MHEG and the Moving Pictures Experts Group (MPEG) to provide the transfer of MHEG objects in MPEG-2 streams, and between MHEG and the Digital Storage Media (DSM) Group to ensure that the DSM Command and Control (DSM-CC) standard provides communication between application servers and end-systems.

Concept

MHEG provides an interchange format for multimedia and hypermedia information and its machine-independent encoding for real-time multimedia applications, as well as for the synchronization and interchange of applications. This includes temporal and spatial relationships between monomedia objects and user interaction mechanisms. MHEG is also a container and description format for each kind of monomedia (e.g. bitmaps, text, video and audio).

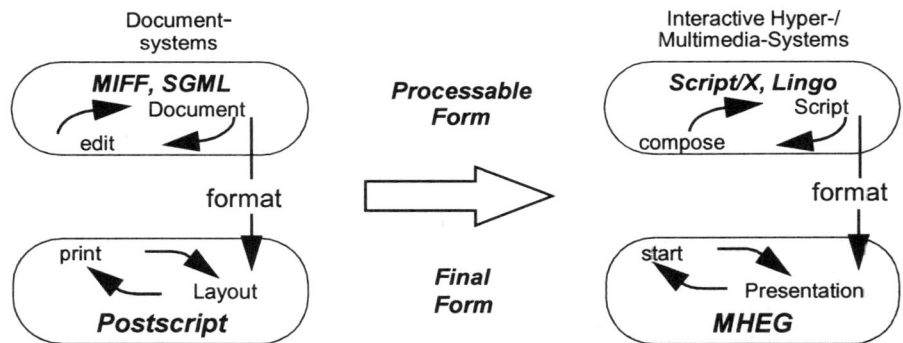

Figure 8: MHEG as Multimedia/Hypermedia Presentation Format

In Figure 8 the analogy between conventional text processing systems and multimedia editors is shown. Analogous to Postscript, the standard page description language for linear documents, MHEG represents the common coding for multimedia and hypermedia applications. MHEG is completely transparent for a multimedia author or designer. Multimedia and hypermedia applications are edited using the same multimedia authoring systems as today. For editing purposes these application are exchanged in a high-level scripting format. When the multimedia authoring process has been finished, the multimedia application is produced by converting the processable-form application to a final-form MHEG application. This layered approach guarantees interoperability between the authoring tools of content providers and the run-time engines in the service consumer systems.

MHEG Classes

MHEG is object-oriented. The MHEG standard defines classes of presentation objects. From the classes, MHEG objects may be instantiated by the presentation designer and interchanged between provider and end-system. This object representation form is called interchanged objects. The following classes are defined in the MHEG-1 standard: Content Class, Multiplexed Content Class, Container Class, Composite Class, Action Class, Link Class, Script Class, Descriptor Class.

4.2.2   HTML and JAVA

HTML is a standard mark-up language and is the basic multimedia description format in the worldwide web domain. HTML was standardized lead by the Internet Engineering Tasks Force (IETF). JAVA defined the language, a byte code, and the virtual machine within an industry-defined standard. A detailed description of the standards is contained in the documents cited.

As a result of the author's prior work, the DAVIC portability group (see [DAVIC 95d]) has shown that both the MHEG and JAVA approaches are suited to implement standardized multimedia applications in an interactive television environment. Given its focus on application and presentation interfaces, MHEG provides a higher-level interface to multimedia programming than JAVA. A detailed comparison of multimedia properties of MHEG and JAVA was given in [Dahm 96].

4.3   Dynamic Generation of Multimedia Applications

Thus far, system and presentation standards for interactive and digital television systems have been introduced. As required in R3 the multimedia application should also be modifiable. This section describes a Virtual Object Store (VOS) which was been designed and implemented to fulfil requirement R3. None of the iTV systems available explicitly addresses the problem of dynamically changing multimedia presentations which require that the presentation structure be created or modified during the presentation run-time.[4]

This book is based on dynamic approaches originating out of work previously done by this author. A brief introduction is given subsequently. A detailed description is given in [Wittig 95b].

---

4. This sections is based on [Wittig 95c].

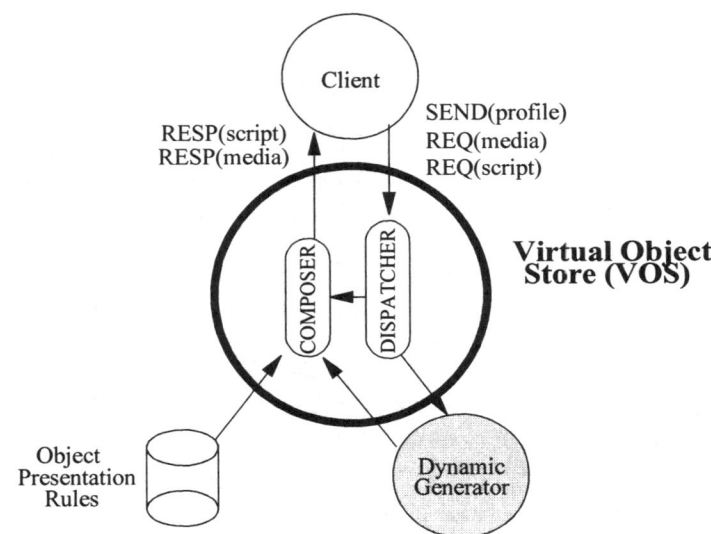

Figure 9: VOS as a Dynamic Presentation Generator

4.3.1   Virtual Object Store

The entity which composes multimedia applications based on multimedia application fragments and a presentation generation rule is called the Virtual Object Store. A multimedia application consists of elementary presentation units which can be dynamically changed when other parts of the application are already active. For example, in a page-oriented multimedia application model such as HTML, all pages which are hyper-linked from the actual page can be changed. However, changes to the actual page are only applied when reloading the complete page. Object-oriented multimedia application models (e.g., MHEG) introduce a flexible mechanism of dynamic generation. For example, time constraints (e.g., duration of an attraction loop) can be dynamically changed though the presentation is already running. Both presentation models are covered by the VOS approach.

Figure 9 depicts the architecture of the VOS. From the iMA client point of view the VOS is a server which delivers multimedia applications or objects on client demand.

The VOS consists of a dispatcher and a composer. The dispatcher reads incoming messages to test whether the client has sent a request for script processing, a request for multimedia application objects, or for information for the iMA. In case a user profile was sent, it is transported to the dynamic generator which is responsible for taking this information into account when delivering the multimedia application objects. Requests for multimedia application objects are given to the dynamic generator which generates and delivers atomic multimedia application objects. These objects are chosen from a list of existing objects (e.g., pointer to a video object) or by dynamically creating the object (e.g., image and text rendering).

The composer is responsible for the reception of atomic multimedia application objects, for the composition of multimedia application objects (e.g., compose a page based on the fragments given by the dynamic generator) and for the delivery of the results to the client.

4.3.2   Location of VOS

The logical server role of the VOS does not imply that the VOS must physically be present on the application server of an iTV system. The VOS can reside on the iTV clients, iTV servers, or partially reside on both. The alternatives are discussed subsequently.

Client-based VOS: In the client-based approach, composition, dispatching and dynamic generation are done at the client site. Because these operations can be resource-intensive, client-based generation presumes powerful client machines. The generation of multimedia applications is required in addition to the multimedia application processing at application run-time.

Server-based VOS: Composition, dispatching and dynamic generation are done at the application server site. Information (e.g., user input or user profiles) is transferred to the server. The structural data which is sent to the client will have already been adapted to the personal requirements of the user. No further modification is necessary.

Mixed Mode: The VOS partially resides on the client and the server. The composer and dispatcher are located in the client node, and the dynamic generator resides on the server.Thereby, complex calculations as required by the application of user profiles are done at the server site while time-critical operations are performed in the clients (without having network delays).

4.3.3 Comparison

Table 3 compares the modes presented above. In the client-based approach, user profiles are kept locally. No additional security functionality is required. However, the client-based approach can result in a high workload for the client system which is critical if small footprint terminals are used as set-top boxes. In the server-based and mixed modes, networks should be protected in order to ensure the privacy of user data. A server-based VOS causes higher server loads. Because functionality is split between server and client in the mixed mode, the load is balanced between the client, server and network.

Table 3: Comparison of Presentation Generation Modes

Criteria	Client-Based VOS	Server-Based VOS	Mixed-Mode
Client Load	⇑a	⇓b	⇒c
Server Load	⇓	⇑	⇒
Network Load	⇓	⇑	⇒
Quality of Network Security	⇓	⇑	⇑

a. The symbol ⇑ specifies a high load or quality.
b. The symbol ⇓ specifies a low load or quality.
c. The symbol ⇒ specifies a medium load.

As result, no specific solution can be generally recommended. The selection of the location of the VOS depends on network bandwidth and computing power of the client and server and these parameters vary according to the run-time environment.

For example, the intelligent media agent's Intranet testbed was comprised of powerful personal computers, a local area network; and a powerful application server. Here, each mode defined in Table 3 was applicable.

5 Architecture of the Intelligent Media Agent

Building intelligent agents requires a multilayered approach (also see [Coen 94]): It is necessary to integrate the existing system interfaces, for example the system software services, the windowing system, the network protocol handling system.[5]

The construction of an intelligent agent requires the integration of middleware functionality in order to provide, among other things, secure transmission, authentication, reliability, simultaneous event and exception handling. Additionally, this middleware functionality is required to implement the computational functions of the intelligent agent.

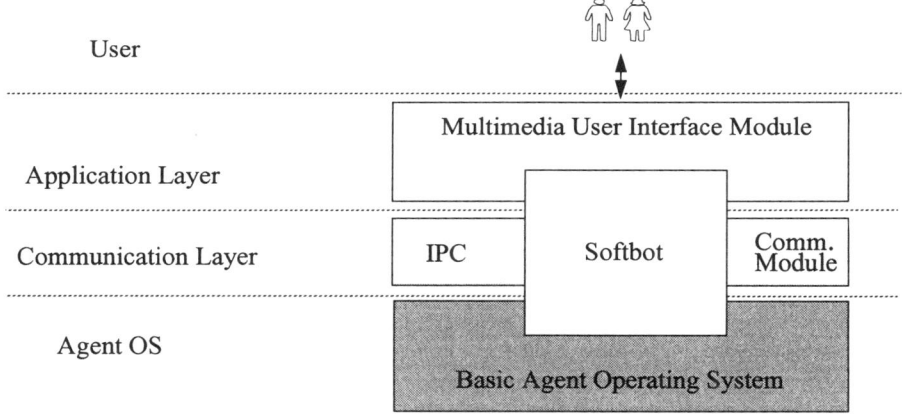

Figure 10: Architectural Skeleton of the Intelligent Media Agent

Many difficulties were found when developing the architecture of the iMA. The architecture presented in these sections is a result of permanent improvements achieved by incorporating the results of previous works (summary in [Brenner 97]) and by considering other works (e.g., [Kautz 94]).

---

5. Parts of this section are published as [Wittig 95c].

5.1 Multilayered Agent Approach

Each of these layers possesses a substantial amount of independent functionality. Therefore, the architecture of the iMA has been designed using a multi-layer agent approach comprised of the modules depicted in Figure 10.

5.2 Module Description

A detailed description of (1) multimedia user interface, (2) high-level softbots, (3) basic agent operating system, and (4) communication environment of the intelligent media agent is contained in the following section.

5.2.1 Multimedia User Interface

The multimedia user interface consists of two parts: the interface to the agent programmer and the interface to the agent user.

The agent programming interface is comprised of components granting access to the agent programming environment and to the interfaces of the basic agent operating system. Typically, this interface is similar to existing programming environments like C++ tools or decision support system programming environments.

The interface to the user of the intelligent media agent is developed using existing multimedia standards in the set-top box already introduced in Section 1 and 2. The personal electronic programming guide is implemented using multimedia presentation standards which developed for the multimedia programming in the set-top boxes of interactive television systems. Additionally, an explanation module is designed as an administration tool which traces the actions of the iMA and explains the results of agent operations.

5.2.2 Basic Agent Operating System

The basic agent operating system is a universal framework for creating and implementing softbots. The basic agent operating system is comprised of an agent implementation language offering high-level primitives for agent knowledge representation and for modelling agent and user behavior, as well as primitives to control the agent activities (e.g., the sharing of an inference engine).

The basic agent operating system implements a run-time environment for agents independent from the specific computational environment in which the agent runs. The basic agent operating system thereby encapsulates the system's specific services and offers general agent programming primitives.

Filtering and Knowledge Acquisition Process

Three major methods for knowledge and meta-knowledge acquisition have been analysed:

(1) Users can specify their interests and the iMA can derive user profiles. Agents can identify user interests by observing the interaction between the user and the iTV system.

(2) The filtering knowledge base contains strategies specifying how the information offered by regarding the user interests should be filtered. For example, a rule could be defined that all programs similar to the interests specified in the user profile be in a personal program.

(3) Agents can apply different strategies to accomplish (1) and (2). They can cooperate by comparing their acquisition strategies, jointly ascertaining the best ways to satisfy user demands. This cooperative component is not considered in this book.

The knowledge acquisition module is responsible for the interpretation of incoming events and uses accumulated histories of user interactions to compute and change user profiles. For example, channel selection events are kept in a history buffer. By evaluating the history buffer agents can recognize user interests. According to requirement R16 three modes of operations are introduced: (1) manual user profile input (detailed specification or by example), (2) automatic user profile generation and adaptation, and (3) a combination of (1) and (2). Details are given in Sections 8.1 and 8.3.

The acquisition strategy is defined by acquisition rules. Each rule consists of a condition and an action list. If the condition is true the rule is put on the agenda of the reference engine. Then all actions in the action list are applied.

The filtering module matches user profiles with TV program listings to identify programs of interest for the individual user. Matching is done by comparing each user profile with each movie profile. The result expresses the similarity between the movie and the profile. For example, a hit list can thereby be formed which contains the best 20 movies for the user. Similar to the acquisition knowledge, the accumulated knowledge of filtering strategies is kept in rules of the filtering knowledge base.

5.2.3 High-level Softbots

Softbots are specific agent entities which are using services offered by the agent operating system. The iMA is comprised of the following softbots: movie softbot, user interface adaptation softbot, interactive services softbot, and the retrieval softbot. These softbots have autonomous inference mechanisms, different techniques for knowledge acquisition and filtering, and different media metadata.

By using inter-agent communication facilities softbots are able to share user profiles. Thereby, all softbots can use user and usage information gained by other softbots. This may result in more information about a specific user and foundation to achieve a higher precision and information consistency in the user model.

5.2.4 Communication Environment

Within the communication environment, service interfaces for distributed and mobile agent communication are required facilitating communication (1) between softbots and the multimedia interface module, (2) internally between softbots, and (3) externally between the agent and its environment including functions for transport of scripts and intelligent agents to basic agent operating systems, their activation and processing, as well as directory and depository services.

Communication with the Multimedia Interface Module

The communication with the multimedia interface module is subdivided into two components: communication with the virtual object server and the explanation module. A so-called Remote Communication Module is responsible for receiving, delegating, and initiating requests from the VOS to the softbots and delivering responses from the softbots to the VOS. Remote inter-process communication facilities, e.g., Remote Procedure Calls (RPC) or socket-based communication classes, are used.

Internal Softbot Communication

In the specific case of the intelligent media agent, softbots need to exchange and share user profiles and knowledge about acquisition and filtering strategies. A softbot of the intelligent media agent does not necessarily know the names and addresses of the other softbots. Based on the discussion of communication systems in [Albayrak 93] two communication models are considered: blackboards and message passing.

The blackboard is a storage which is readable and writable. The metaphor of blackboards is based on [Newell 62] and describes a set of workers/experts who are using a blackboard containing the actual information related to the problem to be solved. The properties of blackboard-based models can be found in [Corkill 91]. Because every message is stored on the blackboard, and agents can autonomously decide to choose this information or not, no addressing scheme for writers and readers is necessary.

Message passing is the transmission of messages from a sender to a receiver. The sender collects the message, specifies the addresses of the receivers, and transmits the message to the receiver by using their specific addresses. Further details are found in [Albayrak 92]. In the broadcast mode each message is sent to all receivers.

Both blackboards and message passing in the broadcast mode are applicable communication models for the intelligent media agent. Because the blackboard approach is a generic model for implementing the communication requirements of softbots, it was chosen for the internal communication among softbots. User profiles and requests are forwarded to the blackboard. The blackboard is a communication module collectively used by all softbots of the intelligent media agent. Incoming information (e.g., interactions, events) is received and stored on the blackboard. Other softbots can read this information from the blackboard and use it for their own tasks. Thereby, the blackboard support softbot communication.

External Agent Communication

An external communication module can be used for the exchange of profiles, as well as acquisition and filtering strategies with other intelligent media agents or as an information broker. This can be done by using standard knowledge interchange formats and languages (Knowledge Interchange Format — KIF [Genesereth 92], Knowledge Manipulation and Query Language — KQML) [Finin 92]). An alternative approach is based on the Common Object Request Broker Architecture (CORBA) which allows the sharing of distributed objects as representations of knowledge. An overview of agent communication methods is given in [Albayrak 93]. KQML and KIF are proposed to implement external agent communication facilities because they include a representation format for knowledge exchange.

5.2.5   Architecture

The resulting architecture of the iMA is shown in detail in Figure 11 with all software components necessary to implement a personal electronic program guide application using a movie softbot.

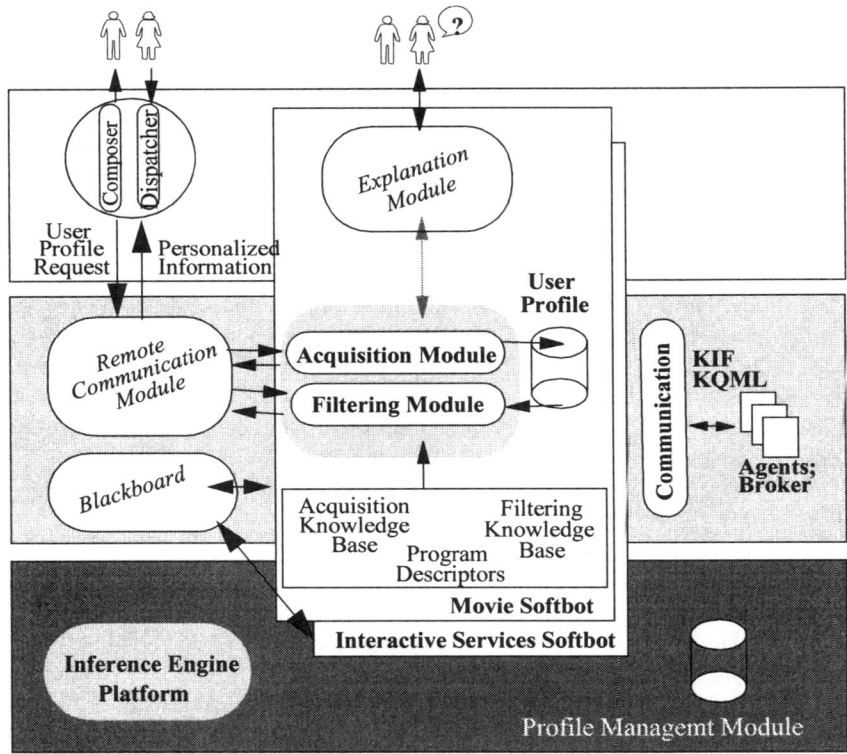

Figure 11: Architecture of the Intelligent Media Agent

5.3   Discussion of Privacy and Security Aspects

Privacy and security issues are important considerations as described in requirement R30. Therefore, the aspect of having an intelligent agent system which keeps user and usage data confidential is an important design goal.

The possibility to offer a secure intelligent agent system depends on the existing agent infrastructure, i.e. (1) the security services of the target client and server environment, (2) the communication between the client and the server modules, and (3) the functionality of the agent modules in the client and server system.

As a general design principle and in order to achieve a minimum level of security, the usage data and user profiles are locally kept in the client system without the ability to access this information via a network. Additionally, access to the local client system is protected by personal identification numbers and passwords.

Due to billing data accumulated for the services used in the interactive television system, it can be assumed that other private data is stored within iTV service provider system. Therefore, the privacy and security issues must be covered within the architecture of the billing system.

6   Program Descriptors and User Profiles

So far, it has been outlined that the intelligent media agent should be based on program descriptors and user profiles. The role of representation and computation has been extensively discussed in the cognitive science research. It has been assumed that main aspects of cognition are based on the internal processing of states which represent states in the external environment. An overview and discussion of user modelling related to artificial life is given in [Wheeler 94]. A user modelling approach is described in [Orwant 96] and implemented as a system called DOPPELGÄNGER. It contains a description of sensors for user metadata, a user metadata model based on Markov models, and clustering as method to group user interests. It is claimed to be application-independent and specializes in defining to user models in computer systems. In the context of the personal EPG, user modelling concentrates on television as an application and the infrared remote control detects user preferences. In addition to the methods described in [Orwant 96], fuzzy attributes are necessary, and it is shown how the user profiles are applied in the context of the personal EPG application.[6]

The following considerations are predicted on the assumption that television programming and user's watching behaviour of it can be structured. The subsequent sections characterize details of iMA components, describing the movie softbot and the personal EPG application. In order to compare user interests and TV programs, two basic data elements of a personal EPG can be identified:

- *Program descriptors* containing information about the TV program.
- *User profiles* containing a specification of user needs, desires and demands.

For this purpose, user profiles supporting viewer preferences have to be installed. Furthermore, administrative procedures are necessary to maintain these personal profiles so that they continually reflect current viewing habits. Otherwise, changes concerning these habits would quickly make them obsolete. The issue of dynamically maintaining user profiles is handled in the following manner: While viewers watch their preferred program, statistical information on channel usage is collected i a heuristic manner. From time to time, the personal EPG uses the accumulated personal access statistics as input parameters for an automatic profile adaptation. The resulting

---

6. This section is based on the diploma thesis [Ehrmantraut 95] which was supervised by the author.

profile is then used to perform the calculation of the new program guide. Hence, no administration is required by the viewer. In the following it is shown how TV information is classified using program descriptors and how user profiles specify individual user interests.

6.1   Classification Units

Channels, programs and scenes are characteristic elementary units of TV transmissions (see Figure 12). The properties of channels, programs and scenes are used to describe the information to be sent. A channel consists of programs (e.g, news, reports, advertisements), each program may have many scenes (e.g., science report, weather, news), and scenes consist of cuts (e.g., weather card, reporter speaking). The classification and description of these data units is a prerequisite for automated information filtering.

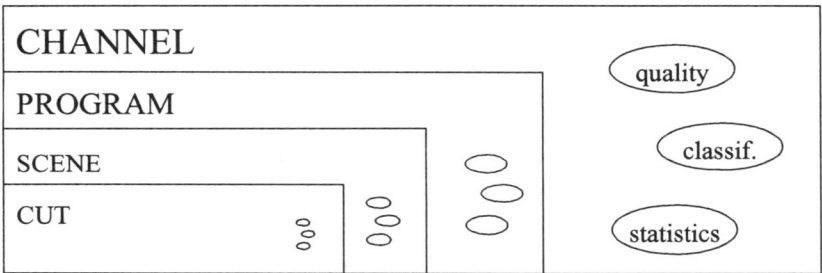

Figure 12: Channel, Program, Scene, Cut

Channels profiles vary from those limited to single-interest programming (e.g., sports, news, music, travel, or home shopping) to those targeting specific age groups (e.g., children television), to those with general programming. For example, the statistical information can show the TV channel's pay mode (e.g., "pay-per-view") or the language (e.g., English).

A program can also be split into smaller units or scenes. Though scenes are generally covered by the classification scheme presented in this section it is not required that personalization be done based on scenes.

6.2   Program Descriptors

A program descriptor may either contain a specification of a TV program or a TV channel. A channel is represented by descriptive data (name, main language, locality, pay mode). For example, these attributes allow such filtering operations as protecting pay-per-view channels with passwords.

The following discussion concentrates on descriptors characterizing TV programs. An analysis of existing printed and digital TV guides yielded the following classes of program attributes:

- Statistics
- Genre
- Quality

6.2.1   Statistics

Table 4 shows the statistical description of the program genres "Show", "Movie" and "Sports" which are based on attributes used in existing movie databases (e.g. Movie Select).

Table 4: Statistical Attributes of Program Descriptors

Program Category	Attributes	Example
Show	- Title - Moderator, producer - Prize - Goal - Participants - Interactivity - Advertisement - Live - Series	100,000 Dollar Show John Little 100,000 Dollars to win games against 3 others yes no yes

Table 4: Statistical Attributes of Program Descriptors

Program Category	Attributes	Example
Movie	- Script author - Producer - Actor(s) - Music - Title - Year of Production - Country - Language - Epoch - Min. age of the viewer - Interactivity, number of branches - Cost - Series - Repetition	King Spielberg Costner Vangelis Dark Christmas 1992 USA English Future 18 (yrs.) 0 2 Dollars no no
Sport s	- Title - Moderator - Importance - Player, opponents - Place, local spectators - Live - Cost	Sports News G. Jauch World Championship Germany vs. Brazil Berlin, Germany yes 5 Dollars

As indicated by the examples, some of the attribute values are of the following types (statistical attributes): integers, real numbers, strings, enumerations. However, a number of attributes have "fuzzy" values such as the epoch a movie is playing in.

There are two classes of statistical program descriptors: (1) general program attributes and (2) provider-specific program attributes.

6.2.1.1 General Program Attributes

Program Epoch

The epoch in which a program is set is an important parameter for program access. For example, children are mainly interested in fictional programs (see [Krüger 96b]) set in the future. Fuzzy specification must be regarded as well as for programs taking place in more than one epoch.

Audio Language

The "Audio Language" attribute specifies the language of the program component. Given multilingual audio transmission facilities in digital television systems, it is possible that users can select their language of choice.

Subtitling

This attribute specifies whether subtitles are transmitted. For example, smaller countries use subtitling as an alternative to audio synchronization. The attribute "Subtitling Language" can be applied.

Subtitling Language

The "Subtitling Language" attribute specifies the language of the program's subtitles.

Repetition

The value "yes" in the attribute "Repetition" indicates that a program is a repeated transmission of a former program. Then the following attributes are applicable:
- *ReferenceDate*
  Specifies the date of the reference transmission
- *ReferenceTime*
  Specifies the time of the reference transmission

Series

A series is a specific program which is split into parts and periodically sent. The following information is added if the program is a series:
- *Sequence Number*
  It specifies the program's number of the serie's number.
- Max. Sequence Number:
  It specifies the maximum number of series. If the overall number is unknown as indicated in the attribute "Continuous Production", the maximum

sequence number contains the maximum known sequence number.

- *Continuous Production:*
  It specifies whether the production of new episodes is in progress/ planned.
- Continuous Story:
  It specifies whether the series contains a continuous story.
- *Overview Available:*
  It specifies the availability of program recaps, i.e. a plot summary of preceeding episodes This attribute is only applicable if the attribute "Continuity" is true.
- *Location:*
  If the attribute "Overview Available" is true the Location specifies the location of the program overview (e.g., as Uniform Resource Locator Reference).

Additional Attributes.

Statistical program attributes can be comprised of additional attributes describing a program's content. Examples:

- number of people killed
- number of goals
- number of music titles in the program

The extraction of these parameters is done by reviewers or automatically. Methods for automatic content classification and description are investigated by researchers on content- based parsing and retrieval (see [Chen 94], [Dimitrova 94], and [Zhang 94]).

6.2.1.2   Provider Specific Program Attributes

Access Path

In broadcast TV system this attribute refers to the TV channel and TV station broadcasting the program. In video-on-demand systems, the access path attribute contains a logical or physical reference to the video.

Date / Time

These attributes specify a program's date and time of transmission. These attributes are not applicable to the video-on-demand mode.

Target Region / Group

This attribute specifies whether a program is of special interest to a specific group or region, these being characterized by descriptors (e.g., zip code).

Age: Preferred and permitted viewers

The parameter "preferred viewers" specifies the age range of the target viewer group. This recommendation is voluntary and may be used to ensure that the viewer can understand this program. The parameter "permitted viewers" specifies the minimum age of viewers permitted to see the program. This parameter is used to restrict program access from viewers younger than the permitted age (e.g., the violent movies).

Advertisements

Most TV networks are free and depend on advertisement for their revenue. In Germany, public and Pay-TV broadcasters require a monthly fee. Depending on country-specific laws and their cost management model they forbid or permit only limited numbers of advertisements to appear (see also Section 1). The parameter "advertisements" specifies whether a program contains breaks for advertisement. If the program contains advertisement additional attributes are applied:

- Number: number of advertisement breaks during the program
- Time: start time of each break
- Length: length of each advertisement break
- Source: source of advertisement (commercial advertisement vs. own TV program previews)

Cost

This attribute specifies the cost of the program. The semantics of this attribute depends on the cost management mode of the TV network (see Section 1).

Preview Mode / Location

The preview mode attribute specifies whether a preview video is available for this program and the reference to the preview video.

6.2.2   Genre

The genre information specifies one or more categories characterizing the content of a program and specifying the type of the television program. An existing approach for the classification of TV programs into genres can be found in the Service Information (SI) section of the Digital Video Broadcasting standard [DVB 94].

6.2.2.1   DVB Service Information

The DVB Service Information standard uses so-called "content identifiers" as program descriptors. These program descriptors are attached to the sent digital programs send in order to provide classification information of television programs. It is usual to define a set of content identifiers for a specific program.

The DVB SI specification consists of the following genres:

- Movie
- News / Current affairs
- Show / Game show
- Sports
- Children / Youth programs
- Music / Ballet / Dance
- Arts / Culture (without music)
- Social / Political issues / Economics
- Education science / Factual topics
- Leisure hobbies

At first glance, this spectrum of genres seems to cover the desired broad range in sufficient depth. However, when analysing the DVB SI description in more detail, the following drawbacks become obvious: (1) just two hierarchy levels are defined in the standard, e. g., the category "Movie" can only be further refined using one level of subclasses; (2) there are only 16 alternatives for each of the two levels of categories. For example, no specific identifiers exist for basketball, golf, or car racing; (3) it is not possible to attach weights to the genres when multiple genres are applied, e.g., in a movie about auto-racing it may be helpful to quantify the dominance of a specific subgenre (Movie or Sports); (4) the information in the classification tree is overloaded, containing information about program attributes like rating information in addition to the basic genre information; and (5) there is no metadata which expresses how the subgenres and parent and child node are related to each other (e.g., usually there is a close relation between thriller and crime movies).

6.2.2.2   Proposed Solution

For these reasons, DVB SI is not as precise as required for the personal EPG. Accordingly, these improvements and refinements are proposed:

(1) A genre tree should contain as many subgenres as are required for the classification of a TV program, see Figure 13.

(2) The number of content identifiers per genre/subgenre be increased to achieve a fine granularity.

(3) The genre tree should contain genre parameters only. For example, the recommended age of the viewer should be extracted from the genre information and stored separately.

(4) As a result of experimental studies based on existing printed TV guides, additional genres and subgenres should be introduced.

(5) Weights should be attached to the edges of the tree specifying the distance between a parent and a child node.

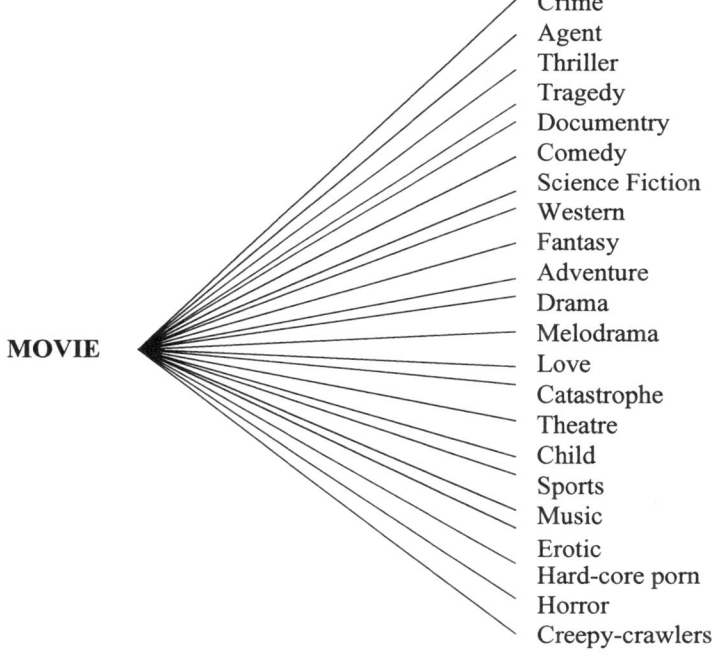

Figure 13: Subgenres of Genre "Movie"

(6) A distance vector should be attached to each child node, specifying the distance from the child node to his siblings.

(7) Priorities should be added to the nodes of the tree in order to express the degree of similarity between the program genre (which is represented by the node) and the program which to be classified.

The improvements (4), (5), (6) and (7) are discussed in the subsequent sections.

Additional Genres and Subgenres

As a consequence of these refinements, the genre tree contains the following top genres:

- Movie
- Information
- Sports
- Music
- Education
- Advertisement
- Show

Each top genre is refined by a sub-tree of subgenres. To illustrate this improvement, Figure 13 and Figure 14 show the sub-trees of the genres "Movie" and "Music" respectively. The genre tree used in the iMA is given in Appendix A.

Weight

In order to specify the distance between a parent and his child, specific weights are assigned to the edges of the genre tree. In the genre tree, each child specifies a subcategory of the parent. The example given in Figure 15 is based on the sports subgenre "fighting sports". In addition to general fighting sports like boxing and wrestling, specific genre was defined for typical asian fighting sports. For example, the subcategory "asian fighting sports" includes judo, karate, and sumo. By specifying this subcategory of fighting sports, the program classification can be simplified. The weight of the edges are defined as depicted in Figure 15. The edge from fighting sports to asian fighting was assigned a value of 0.1 indicating that asian fighting is more related to the genre fighting sports than to boxing or fencing. Hence, the distance between "fighting sports -> boxing" and "fighting sports -> judo" is similar.

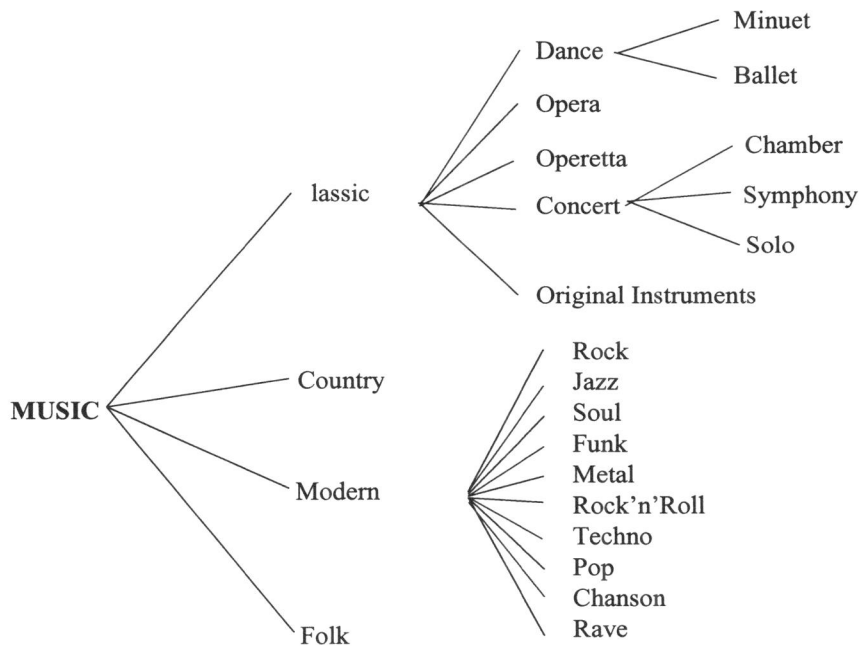

Figure 14: Subgenres of Genre "Music"

Distance Vector

A distance vector is attached to each child node. This vector specifies the distance of this child node to his siblings and can be used to compare attributes which are in different sub-trees but share a common antecedent. The greater the distance the higher are the values of the elements in the distance vector. An example is shown in Figure 16, where it was specified that boxing is more related to asian fighting sports than to fencing.

Discussion of Priorities

Based on the genre tree, a TV program can be classified by enumerating the characteristic categories (enumeration method). For example, the movie "Terminator" is a science fiction movie also containing minor love story elements. By using the

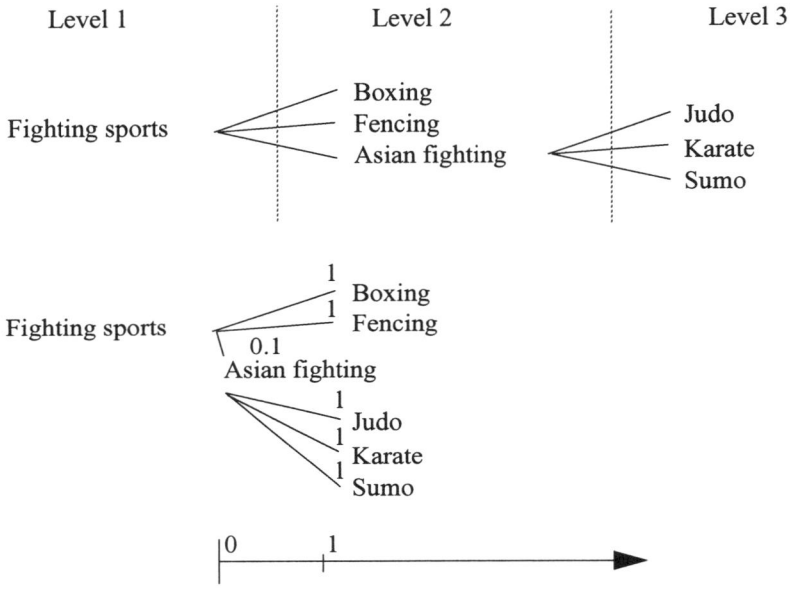

Figure 15: Weights of Edges in the Genre Tree

enumeration method "Terminator" is classified as MOVIE->Science Fiction and MOVIE-> Love. However, it is not possible to specify that the subgenre "Science Fiction" is more significant than the "Love" one.

The prioritized enumeration method defines that more significant genre information is listed before less significant genre information. The "Terminator" program contains (MOVIE -> Science Fiction; MOVIE-> Love). Though priorities can be expressed it cannot be specified to what degree a subgenre dominates.

Therefore, a parameter "priority" is attached to each node in the genre tree. The priority of a node quantifies the significance (in %) related to the whole program genre specification. For example, the program "Terminator" is rated (MOVIE->Science Fiction, 95%; MOVIE->Love 5%). Thereby, all subgenres in a program can precisely be assigned.

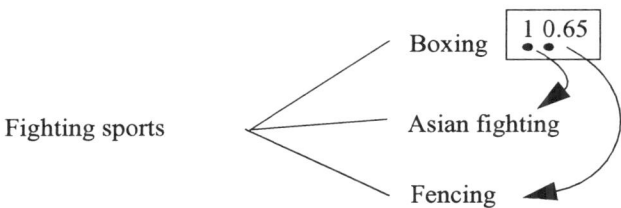

Figure 16: Example Distance Vector

Problems

An objective assessment of program genres is not possible and therefore subject to discussion. It is expected that more than one classification exists for the same program. A user prefers classification schemes which reflect his personal opinion. The definition of the genre tree is not standardized. For defining program and user profiles, a standard format for genre trees is required and only partly exists (e.g., as defined by the DVB SI standardization group). In addition, it is necessary to assign weights to the tree's nodes.

Because new internet-based TV networks and programs will emerge in the next decade, extensions of the genre tree can be expected. There is a direct correlation between the precision of the classification into genres and the effort to classify the program. By using the weighted genre trees the classification effort is scalable.

6.2.2.3   Quality

Printed TV guides contain qualitative information, e.g. ranking lists, tips of the day, or reviewer annotations. Though these parameters are not objective, they may help users as additional assessment criteria. To capture this kind of information, existing TV guides were analysed. As a result, the TV program is qualitatively assessed in terms of its:

- suspense
- humor
- action
- sexual content

The attribute values are proposed to range from 0% to 100%. A value of 100% means that the attribute is applicable, 0% means not applicable.

6.3   User Profile

So far, the program description structure for categorizing television programming has been outlined which embodies the various program offers in the intelligent media agent system. In order to allow users to specify their program requirements (see R15) and to accomplish an automatic matching of user preferences with the best available offer, user profiles are needed.

6.3.1   Identification and Privacy Aspects

User identification is accomplished by using end-system identifiers, personal identification numbers (PINs), or by service or credit card infos. Because the iMA not expected to have identification methods for each individual user, it also manages a group profile. In this group mode, members of the group are considered to act as one virtual person. A refinement of this approach to manage individual group members based on a-priori knowledge is underway.

User profiles contain private, that is, sensitive information. Therefore, protection against unauthorized access is a key requirement. Privacy is achieved by executing the iMA system in the local set-top box without granting outside read access to the user profiles, or by keeping the profiles in a confidential application server and guaranteeing secure transmission between application servers and end-systems.

6.3.2   Desires and Demands

The strength of a user requirement can vary. A desire characterizes a weak recommendation only and does not necessarily need to be regarded when determining favourite programs. A user demand, however, is a strong requirement and needs to be regarded when assembling programs for the guide. Desires and demands can be positive or negative. This results in the classification is depicted in Figure 17:

1. *Positive Desires:* The user recommends program categories. As this specification is only a desire, it does not need to be necessarily regarded by the iMA.
2. *Negative Desires:* The user specifies program categories which are not of interest for him. This specification does not need to be necessarily regarded by the iMA.
3. *Positive Demands:* Hereby, the user instructs the iMA to regard all programs satisfying the specified demand. Demands cannot be ignored.

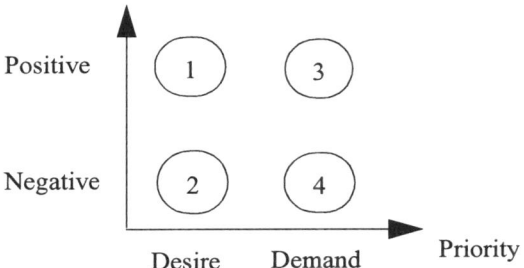

Figure 17: Classification of Desire and Demands

4. *Negative Demands:* The user instructs the iMA to ignore all programs matching the user specification. These instructions cannot be ignored.

Wishes, likes, dislikes, desires and demands do often appear to be fuzzy. For example, there are steps in between weak/strong desires and a weak/strong demands. Therefore, in addition to the model presented in Figure 17, fuzzy sets are proposed to describe user profiles.

6.3.3 Time of Day, Day of Week, Holiday

Because the preferences are dependent on the time of day, day of week, and holiday schedules (see requirement R14) these parameters should be attached to each user profile.

Time of Day

The relevant time period for a user's television program interest is defined by adding two parameters containing the start- and end-points to his user profile.

Day of Week

The parameter "day of week" is added to the data definition of a user profile in order to specify which days of a week a user profile is valid.

Holiday

This parameter lists the holidays for which the user profile is valid.

6.3.4   Fuzzy Attributes

When defining the user profile, similarities between program descriptors and user profiles can be found: The interest which should be expressed in the user profile mainly refers to a specific program, a program attribute, or a set of programs which characterize the interest profile for an individual user. Therefore, the format of user profiles is a superset or subset of program descriptors.

It has been defined that user profiles should also reflect the priorities. Additionally, for each attribute, information about the priority is contained in the data definition of user profiles. This priority information is used to reflect the importance of specific attributes in the overall profile specification. A value of zero means that the user does not care about the values in the user profile attributes. If the priority is 100, the attribute is of great importance in the matching process. The comparison of two values depends on the priority of the attribute. Accordingly, the result of a comparison of two fuzzy sets is a fuzzy result. In the following, examples of the priority and fuzzy specification of user attributes are described.

Example 1: A user likes movies with much humour independent of the actual actors. This is expressed in his user profile by giving the genre attribute "humour" a high priority of 90 and setting the "actor's" attribute priority to zero.

Example 2: A user prefers early rock and roll movies. This is expressed by assigning the music subgenre "rock and roll" a high priority and a high value, and by describing the movie's era of the movie with the fuzzy description "Sixties".

Example 3: A user can only speak English. Therefore, the iMA is instructed to ignore all other languages by assigning the value "English" to the attribute "languages" in his user profile.

6.4   Attribute Types

Four elementary attribute types are used to store attributes:

- *Percent Attribute Type (PAT):* It stores floating point numbers between 0 and 100 (i.e., quality in percent). When calculating the average of multiple PAT's, the result is of the same type.

- *Number Attribute Type (NAT):* To store numbers (i.e., producing year) a NAT is required. It has the same characteristics as the PAT when calculating the average. The main difference of these two types is their scope.

- *Discrete Attribute Type (DAT):* With DAT text strings can be stored. The comparison of DAT's is a boolean result (equal or not equal).

- *Hierarchy Attribute Type (HAT)*: This attribute type expresses hierarchical dependencies between values of an attribute (e.g., genre trees).

The above types can be extended to *Multi-Attribute Types* in order to store more than one value for each attribute. The MAT contains additional attributes which specify the share of every elementary attribute type in the MAT. From a formal point of view, MAT is defined as the union of PAT, NAT, DAT and HAT:

$$MAT = PAT \cup NAT \cup DAT \cup HAT$$

For example, the attribute "Genre" (see Section 6.2.2) is a MAT containing NAT's and HAT's. The attribute "actors" is a MAT consisting of DAT's.

7 Matching and Clustering

Having described the structure and meaning of program descriptors and user profiles, matching and clustering algorithms applied to the program descriptors and user profiles are explained in this section.[7]

Related Work

A large amount of related work (e.g., [Salton 87], [Sheth 94]) can be found in the area of information and document retrieval research and information filtering research in the Internet. Based on this work fundamental matching and clustering algorithms have been developed. The original approach consists of the application of these elementary algorithms to the fuzzy sets presented in the previous section.

7.1 Matching Methods

The comparison of program descriptors and user profiles - both containing fuzzy values - is the basic operation in the program filtering process. The desired result of this matching or filtering operation is an optimization of the personal program guide, the "best" programs, i.e. the one's must closely matching to a user's interests. For this purpose, a program profile and a user profile are compared by using the so-called MATCH function. The result of this function indicates the degree of similarity between the user and program profiles.

7.1.1 Definition

Program descriptors and user profiles are data structures containing sets of attributes. In each set, the type of attributes may be different. The definition of the MATCH function is as follows:

$$\text{MATCH: } R^{n+1} \to R \text{ with}$$

$$\text{MATCH(V)} = \frac{|V|}{|\xi|} \qquad \text{Def. 1}$$

---

7. This section is based on the diploma thesis [Ehrmantraut 95] which was supervised by the author. Parts have been published as [Ehrmantraut 96].

$$\text{with:} \qquad \xi = \begin{bmatrix} \xi_0 \\ \xi_1 \\ \dots \\ \xi_n \end{bmatrix} = \begin{bmatrix} 1 \\ 1 \\ \dots \\ 1 \end{bmatrix} \Rightarrow |\xi| = \sqrt{n+1}$$

where V is a vector containing the attributes of a program descriptor or a user profile. $V \subseteq R^n$ and $|x|$ are norms to the function MATCH (e.g., Euclidian norm).

The result of MATCH indicates the degree of similarity between program descriptor and user profile. The components of the vector V and the result of MATCH are in the interval [0, 100]. For example, a result with 100 as value indicates that program descriptor and user profile are identical. If the result of the match function value is zero then there is no similarity between the program descriptor and the user profile.

The vector V is the result of sub-function MATCH_v. The vector V contains the result values of an attribute-to-attribute comparison of complete profiles. $V = (v_0, v_1, \dots, v_i)$ where i is the maximum number of attributes in the profile where

$$|V| = \sqrt{\sum_i (v_i)^2} \qquad \qquad \text{Def. 2}$$

$$\text{MATCH}_v : PB \times P \times W \rightarrow R^{n+1} \text{ where}$$

$\quad$ P : Program profile; $P \subseteq MAT^{n+1}$
$\quad$ B : User profile; $B \subseteq MAT^{n+1}$
$\quad$ PB : $P \cup B$
$\quad$ W : Similarity vector; $W \subseteq R^{n+1}$
$\quad$ R^{n+1}: Result vector

$$\text{MATCH}_v (pb, p, w) := \text{MATCH}_A (pb_i, p_i, w_i) \text{ where}$$

$\quad$ A is the attribute type of pb_i and p_i;
$\qquad\qquad A \in \{NAT, HAT, DAT, PAT\}$
$\quad pb = (pb_0, \dots, pb_n) \qquad pb \in PB$
$\quad p = (p_0, \dots, p_n) \qquad\quad p \in P$
$\quad w = (w_0, \dots, w_n) \qquad\quad w \in W$

and $\qquad$ $\text{MATCH}_A : X \times Y \times G \rightarrow R$

$\text{pb}_i \in X$
$X \in \{\text{NAT}, \text{HAT}, \text{DAT}, \text{PAT}\}$

$p_i \in Y$
$Y \in \{\text{NAT}, \text{HAT}, \text{DAT}, \text{PAT}\}$

$w_i \in G; G \in R; \text{ and } 0 \leq w_i \leq 100$
$i, n \in N \text{ and } 0 \leq i \leq n$

Since there are multiple attribute types A, specific match functions MATCH_A for each attribute type are required. Hence, we need a number of comparison functions like those defined in the function MATCH_A. For example, MATCH_A compares the values of two similar attributes in the program and user profiles. Additional methods are shown below.

7.1.2 Arithmetical Difference Method

The arithmetical difference method is used to calculate the similarity of two vectors. The result is the sum of the positive difference vector. The type specific function MATCH_A calculates the difference of each vector element as follows:

$$\text{MATCH}_A : X \times Y \times G \rightarrow R$$

$$X \in \{\text{NAT}, \text{HAT}, \text{DAT}, \text{PAT}\}$$
$$Y \in \{\text{NAT}, \text{HAT}, \text{DAT}, \text{PAT}\}$$
$$G \in R$$

$$\text{MATCH}_A(x, y, g) := \begin{cases} 100 & \text{if } |x - y| < f_A(g) \\ \\ 0 & \text{else} \end{cases} \qquad \text{Def. 3}$$

$$x \in X$$
$$y \in Y$$
$$g \in G$$

f_A (g) is a threshold which is a function of attribute priority g and the attribute type A. An example function f_A (g) is given in Figure 18, where f_{max} is defined to be the maximum value of the attribute A. For example, the maximum f_{max} of the PAT (Percent Attribute Type) is 100%.

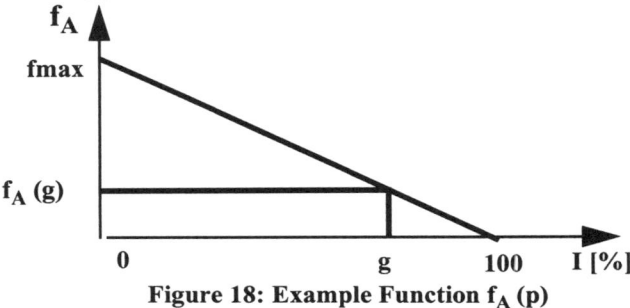

Figure 18: Example Function f_A (p)

The arithmetical difference method is applicable to all attributes with difference ("-") operations defined. One drawback of this method is the lack of fine granularity in the results (only 0 or 100). An extension is required.

7.1.3   Fuzzy Set Matching Method

The fuzzy set matching method allows the comparison of fuzzy attribute sets. Attributes of the program description, the user profile and the priority of the user profile are input parameters of the fuzzy set matching method. The fuzzy set matching function is defined by:

$$\text{MATCH}_{AF} : X \times Y \times G \rightarrow R$$

$$\text{MATCH}_{AF} (x, y, g) := f_{Axg} (y)$$

Figure 19 shows an example function f_{AF} for an attribute A.

The result of this function depends on the priority g which is used to set the fuzzy parameters t_1, t_2, t_3 and t_4. Hence, each attribute may have different set of fuzzy parameters. Based on this definition, the similarity between a given pair of attributes can be measured. If $t_1 = t_4$ and $t_2 = t_3$ the fuzzy set is symmetric, otherwise asymmetric.

By assigning specific parameters to the attributes, the ability to model relations between attributes of different types is gained.

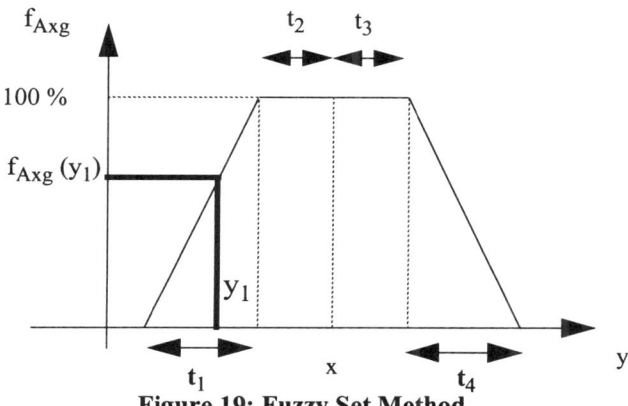

Figure 19: Fuzzy Set Method

7.2 Application to Attribute Types

Thusfar, a fuzzy set matching method has been defined. This method is refined in the next section and applied to elementary attribute types as already cited in Section 6.4.

7.2.1 Percent Attribute Type (PAT)

The values of percent attributes range from 0% to 100%. As mentioned above, the parameters t_1, t_2, t_3 and t_4 should depend on their attribute priority g. A lower priority means that the similarity of this attribute is less important. Additionally, a parameter c is introduced:

$$t_i := \overline{g} * c_i \qquad , c_i \text{ .. parameter for the calculation of } t_i \qquad \text{Def. 4}$$

$$\overline{g} = 100\% - g \qquad \text{Def. 5}$$

The PAT fuzzy set matching method is similar to the function shown in Figure 19. The following auxiliary variables are introduced resulting in the PAT fuzzy set as depicted

$$X_{ll} := x - (t_1 + t_2) \qquad X_l := x - t_2$$
$$X_{rr} := x + (t_3 + t_4) \qquad X_r := x + t_3 \qquad \text{Def. 6}$$

in Figure 20.

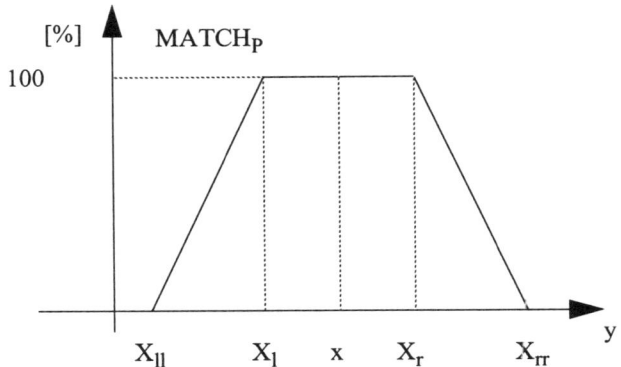

Figure 20: PAT Fuzzy Set Model

$MATCH_{PAT}$ is defined as follows:

$$MATCH_{PAT}(x,y,g) := \begin{cases} 0 & \text{if } (y \le X_{ll}) \text{ or } (y \ge X_{rr}); \\ (y-X_{ll}) * 100 / t_1 & \text{if } (X_{ll} < y < X_l); \\ (y-X_{rr}) * 100 / -t_4 & \text{if } (X_r < y < X_{rr}); \\ 100 & \text{else.} \end{cases} \qquad \text{Def. 7}$$

In Figure 21, Figure 22, and Figure 23, three examples of PAT fuzzy sets are shown based on the parameters $c_1 = c_2 = c_3 = c_4 = 0.5$, the attribute value x=60%, and the three priority values 100%, 60% and 0%.

In Figure 21 the result of the match operation is 100% if the value is identical to the attribute value x. Otherwise, the result is 0. X_{ll}, X_l,X_r, and X_{rr} are set to 100%.

In Figure 22 the attribute priority g is important (i.e., $g = 60\%$). The auxiliary parameters are X_{ll}=20%, X_l= 40%, X_r= 80%, and X_{rr}=100%. The result of the matching operation is 100% if the values are in the interval [40,80].

Figure 23 shows an attribute which is unimportant. Its priority g is set to 0%. The auxiliary parameters are X_{ll}=0%, X_l= 0%, X_r= 100%, and X_{rr}=100%. The result of the matching operation is always 100%.

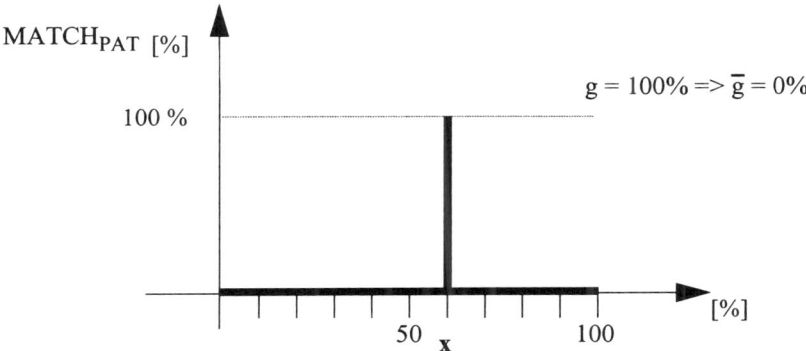

Figure 21: Example 1 "critical" (g = 100 %)

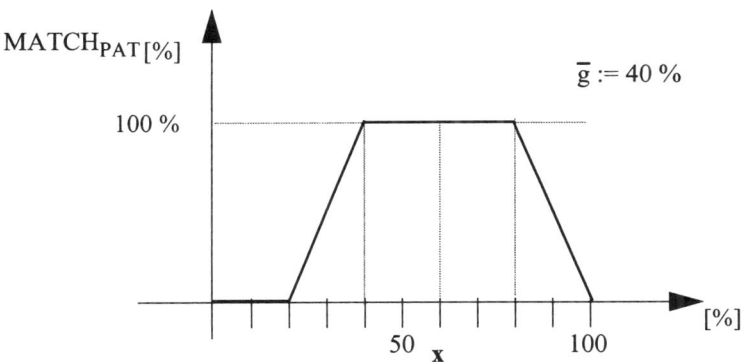

Figure 22: Example 2 "important" (g = 60 %)

It can be seen that the priority g is an important parameter for defining the fuzzy set of an attribute and for matching Percent Attribute Types.

7.2.2 Number Attribute Type (NAT)

The fuzzy set matching method of NATs is similar to that of the PAT's. The factors c_1 to c_4 are adapted, and the x-axis covers another range.

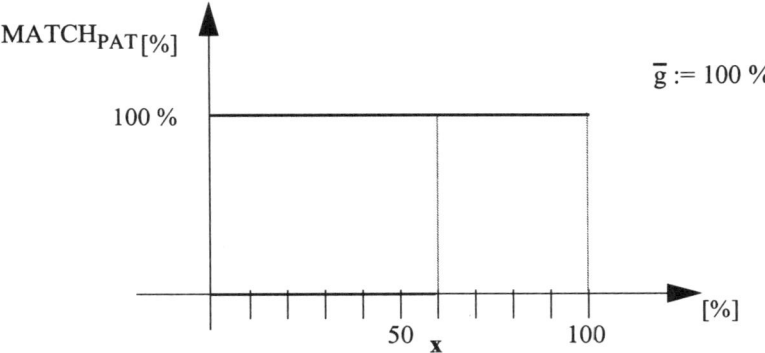

Figure 23: Example 3 "indifferent" (g = 0 %)

By taking the NAT "production year" as example, Figure 24 illustrates the fuzzy set matching method of NAT's. The fuzzy set parameters are as follows: $c_1 = c_2 = c_3 = c_4 = 0.25$. The parameters of the fuzzy set are adapted to the range of the specific attribute.

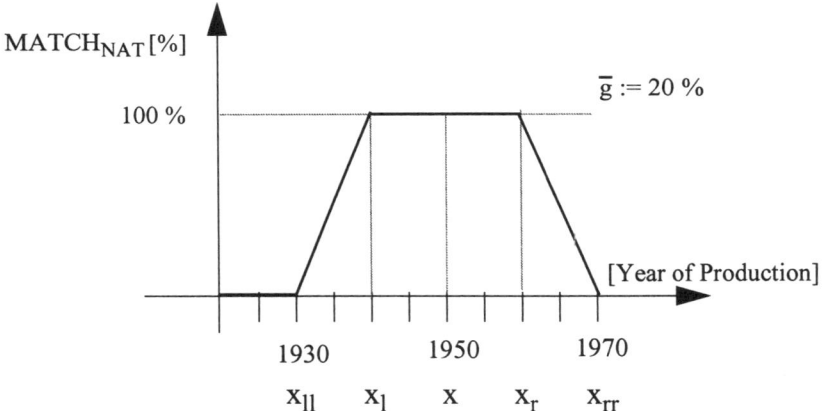

Figure 24: Example: Year of Production x = 1950 ; g = 80 %

$MATCH_{NAT}$ is defined as follows:

$$MATCH_{NAT}(x, y, g) := MATCH_{PAT}(x, y, g) \qquad \text{Def. 8}$$

7.2.3   Discrete Attribute Type (DAT)

In the context of the iMA, there is no order among the DAT values. The matching function of DATs is defined as follows:

$$\text{MATCH}_{\text{DAT}}(x,y,g) := \begin{cases} 100 & \text{if } x = y \\ \bar{g} & \text{else} \end{cases} \qquad \text{Def. 9}$$

When matching equal attribute values the result is always 100%. If the attributes are different, the lower is the priority g of this attribute the higher is the matching result. For example, if g=0% ("indifferent") the result is 100%. This is motivated by the fact that attributes which are unimportant should not restrict the matching result of the whole attribute set.

7.2.4   Hierarchy Attribute Type (HAT)

In addition to the matching methods presented for DATs, the relations between the attributes in the hierarchy must be regarded. The following specific cases can be identified:

1.   The result of matching equal attribute values should be 100%.
2.   If an attribute is a root node of the other attribute, the similarity depends on the weights assigned to the edges between both nodes. The result of the matching operation should increase with a decreasing distance between both attributes.
3.   If both attributes belong to different sub-trees, and they have a common root node, then the result of the matching operations depends on the values attached to the edges between the attributes and their sub-tree children of the root nodes, and the distance between them.

A distance function d is introduced which determines the distance between two nodes in the HAT tree. If $w(E_{x,y})$ is defined as the weight of the edge between attributes X and Y and $dv(X,Y)$ is the distance vector between two children X and Y of the same parent, then d (A,B) is defined as follows:

$$
d\,(A,B) := \begin{cases}
0 & \text{if } A = B; & \text{Def. 1} \\
d(A,C)+w(E_{C,B}) & \text{if C is parent of B, and B belongs} \\
& \text{to a subtree with A as root node;} \\
d(C,B)+w(E_{C,A}) & \text{if C is parent of A, and A belongs} \\
& \text{to a subtree with B as root node;} \\
d(V,A)+d(W,B)+1 & \text{if A and B have C as common} \\
& \text{root node, V is child of C in the} \\
& \text{same subtree as A, and W is child} \\
& \text{of C in the same subtree as B.}
\end{cases}
$$

In order to map the result of the distance function d into the fixed interval (0,1) which is required by the definition of MATCH function in Section 7.1, the function f is introduced:

$$
f(A,B) := g^{|d(A,B)|} \qquad\qquad \text{Def. 11}
$$

Additionally, a threshold T is introduced in Def. 12 which ends the calculation of $d(A,B)$ as soon as the distance T was exceeded. Hence, MATCH_{HAT} is defined as follows:

$$
\text{MATCH}_{HAT}\,(A,B) := \begin{cases}
f(A,B) & \text{if }(A,B)| < T; \\
\overline{g} & \text{else.}
\end{cases} \qquad \text{Def. 1}
$$

Example:

The example genre tree is given with $g = 0.95$ in Figure 25. The result of MATCH_{HAT} ("sports", "karate", 5%) is:

$$
d(\text{"sport"}, \text{"karate"}) = 2.1
$$
$$
\text{MATCH}_{HAT} = f(\text{"sports"}, \text{"karate"}, 5\%) = 0.95^{2.1} = 0.89
$$

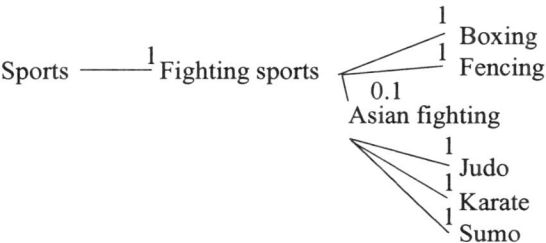

Figure 25: Example of a HAT Genre Tree

7.2.5 Multi-Attribute Types (MAT)

Elementary attribute types can be aggregated into multi-attribute types. Two multi-attribute types can be compared by applying a specific MATCH function to the the results of the match operation of the corresponding elementary attributes:

Attribute A = { (A_1 , a_1) , (A_2 , a_2) , ... , (A_n , a_n) }
Attribute B = { (B_1 , b_1) , (B_2 , b_2) , ... , (B_m , b_m) }

$m \in N; n \in N$
$A_n \in \{NAT, HAT, DAT, PAT\}$
$B_m \in \{NAT, HAT, DAT, PAT\}$
$a_n \in PAT; b_m \in PAT$

The functions t_g and t_a which compare the corresponding tupel (A_i,B_i) and (a_i,b_i) are defined as:

$t_{AB} (i , j , g) := MATCH_A (A_i , B_j , g)$

$i \in N; j \in N$
$A_i \in \{NAT, HAT, DAT, PAT\}$
$B_j \in \{NAT, HAT, DAT, PAT\}$
$g \in G; G \in R;$ and $0 \le g \le 100$

$t_{ab} (i , j , g) := MATCH_{PAT} (a_i , b_j , g)$

$i \in N; j \in N$
$a_i \in PAT; a_j \in PAT$
$g \in G; G \in R;$ and $0 \le g \le 100$

The similarity of (A_i,a_i) and (B_i,b_i) is defined by multiplying both similarities t_{AB} and t_{ab} (see Def. 13). A threshold for the minimum result of t_{ab} is defined.

$$t(i,j,g) := \begin{cases} t_{AB}(i,j,g) * t_{ab}(i,j,g) & \text{if } t_{ab}(i,j,g) > T_g; \\ \overline{g} & \text{else.} \end{cases} \qquad \text{Def. 13}$$

By using the function $t(i,j,g)$ each attribute (A_i,a_i) of A is matched with the corresponding attribute (B_i,b_i) in B. Because it is useful to detect attributes in the MAT having a maximum similarity, the following formulae are applied. It is assumed that $n < m$.

$$Z := \{ t(i,j,g) \mid \quad \begin{aligned} & i \in N; j \in N, 0 \leq i \leq n, \ 0 \leq j \leq m \} \\ & g \in G; G \in R; \text{ and } 0 \leq g \leq 100 \\ & |Z| = z = n*m; \end{aligned} \qquad \text{Def. 14}$$

A function MAX is introduced which calculates the n maximum values of set G and returns them as set M.

$$MAX(M,G,n) := \begin{cases} M & \text{if } n=0; \\ MAX(M \cup \{max(G)\}, G \setminus \{max(G)\}, n-1) & \text{else.} \end{cases}$$

The result of $max(G)$ is the maximum value of set G.

The result of Max $(Z,n) := MAX(\{\},Z,n)$ is a set containing a number of n maximum values of the corresponding attributes in set A and B. MATCH$_{MAT}$ applies the arithmetic average to these results. Its definition is given in Def. 16:

$$Sum(M) := \begin{cases} 0 & \text{if } M = \varnothing \\ m + Sum(M \setminus \{m\}); m \in M & \text{else} \end{cases}$$

$$\text{Def. 16}$$

$$MATCH_{MAT}(A,B) := \frac{Sum(Max(Z,n))}{n}$$

7.3   Matching Example

So far the different types of matching operations have been specified for the PAT, NAT, DAT, MAT and HAT. The subsequent section gives a detailed example which shows the matching of two movies.

Movie Attribute

The example movies are defined in the following table:

Table 5: Example Movies

Attributes	Attribute Type	A	B
Name	DAT	The Killer	The Funny Killer
Director	DAT	Roger Rabbit	Donald Duck
Year of Production	NAT	1985	1992
Suspense	PAT	90%	70%
Genre	MAT	(Crime, 100%)	(Crime, 60%) (Comedy, 40%)

The genre tree which is relevant in this context is depicted in Figure 26. The priority g of all attributes is set to 90%.

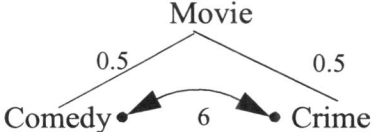

Figure 26: Genre Subtree

MATCH (A , B , 90%)

$g = 90\% \implies \bar{g} = 10\%$ acc. to Def. 6

1) $MATCH_{DAT}$ ("The Killer", "The Funny Killer", 90%) = 10% acc. to Def. 9

2) $MATCH_{DAT}$ ("Roger Rabbit", "Donald Duck", 90%) = 10%   acc. to Def. 9

3) $MATCH_{NAT}$ (1985 , 1992, 90%) = 60%   acc. to Def. 8
    $c_i = 50 \Longrightarrow t_i = 0.1 * 50 = 5$(acc. to Def. 4)
    (see also Figure 27)

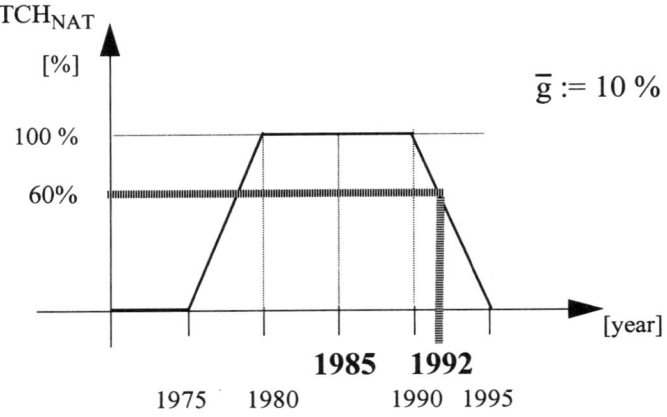

Figure 27: $MATCH_{NAT}$(1985, 1992, 90%)

4) $MATCH_{PAT}$ (90% , 70%, 90%) = 50%   acc. to Def. 7
    $c_1 = c_4 = 2 \Longrightarrow t_1 = t_4 = 0.1 * 2 = 20\%$
    $c_2 = c_3 = 1 \Longrightarrow t_2 = t_3 = 0.1 * 1 = 10\%$
    (see also Figure 28)

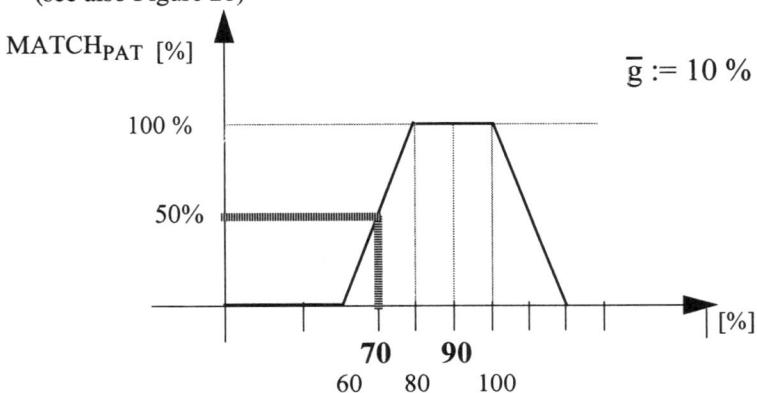

Figure 28: $MATCH_{PAT}$ (90% , 70%, 90%)

5) $\text{MATCH}_{\text{MAT}}$ (G1, G2):
G1= { (Crime,100%) }   n= |G1| = 1
G2= { (Crime,60%) , (Comedy,40%)}   m= |G2| = 2
T = 10%

Z = { $\text{MATCH}_{\text{HAT}}$ (Crime, Crime, 90%) * $\text{MATCH}_{\text{PAT}}$ (100% , 60%, 90%) ,
      $\text{MATCH}_{\text{HAT}}$ (Crime, Comedy, 90%) * $\text{MATCH}_{\text{PAT}}$ (100% , 40%, 90%) }
                                                                    acc. to Def. 14

5.1) $\text{MATCH}_{\text{HAT}}$ (Crime ,Crime , 90%) = 100%   acc. to Def. 12
5.2) $\text{MATCH}_{\text{PAT}}$ (100% , 60% , 90%) = 50%   acc. to Def. 7
5.3) $\text{MATCH}_{\text{HAT}}$ (Crime , Comedy , 90%) = 10%   acc. to Def. 12
5.4) $\text{MATCH}_{\text{PAT}}$ (100% , 40% , 90%) =10%   acc. to Def. 7

$\Rightarrow$ Z = { 50% , 10 % }   acc.to Def. 14
$\Rightarrow$ MAX (Z , 1) = { 50% }   acc.to Def. 15
$\Rightarrow$ $\text{MATCH}_{\text{MAT}}$ ({Crime,100%},{Crime, 60%;Comedy,40%}) = 50%
                                                                    acc.to Def. 16

$$V := \text{MATCH}_{\text{v}} \ (\ A \ , B \ , G \ ) = \begin{bmatrix} 10\% \\ 10\% \\ 60\% \\ 50\% \\ 50\% \end{bmatrix} \Rightarrow |\xi| = \sqrt{5} \quad \text{acc. to Def. 1}$$

$$\|V\| = \sqrt{0{,}1^2 + 0{,}1^2 + 0{,}6^2 + 0{,}5^2 + 0{,}5^2} = \sqrt{0{,}88} = 0{,}938$$

$$\text{MATCH} \ (\ V) = \qquad\qquad\qquad \text{acc. to Def.}$$

The approximate similarity of the example programs is 42%.

7.4   Clustering

Clustering methods are the basic algorithms used for the adaptation of profiles. In the context of the Personal EPG, application clustering methods are used to interpret the feedback given by users when watching TV programs. Similar usage profiles have to be clustered, new interests detected, new sub-profiles merged with existing profiles, old interests removed and existing profiles updated.[8]

In the context of the iMA, clustering methods combine a set of profiles into a superset containing similar elements. The clustering methods used in this book were derived from clustering methods used in document retrieval systems listed below:

Clustering methods are divided into non-hierarchical and hierarchical methods. The non-hierarchical ones typically have a low complexity, e. g. $O(n)$ or $O(n \log n)$ [Willet 88], which leads to the disadvantage that their results depend on the sequence of the elements processed. In the context of document retrieval, precise results are more important than computation time [Rasmussen 93]. Therefore, most document retrieval systems use hierarchical, agglomerative clustering methods (HACM). In the following, a selected list of HACM's is presented. A detailed discussion of HACM is given in [Henrion 88] and [Mucha 92].

- The single-linkage-method [Jarvis 73] groups elements with their nearest neighbors. Their essential characteristics are high stability and a tendency to produce chains of elements (chaining effect). Such a property includes a risk of establishing large clusters.

- Contrary to the single-linkage-method, the complete-linkage-method calculates the elements with the lowest similarity (Furthest-Neighbor-method). Usually, the result is a set of many small clusters.

- The average-linkage-method produces clusters that contain documents which have a higher average similarity to the remaining members of their cluster than to all other documents in other clusters [El-Ham 89].

- Ward's method produces clusters based on the standard deviation. The criterion is a minimum overall standard deviation in the clusters.

---

8. Parts of this section are published in [Michel 97] which was supervised by the author.

High stability and high precise results are requirements (see R24 and R25) applied to the Personal EPG clustering method. Previous research results concerning clustering approaches (see [El-Ham 89]) indicated that the single-linkage-method clustering algorithm is best-suited to meet the stability and precision criteria. Ward's method is a conservative approach which is recommended if contractions in the coordinate system should be avoided (see [Henrion 88]).

Figure 29 shows some sample clusters of two-dimensional profiles where the characters symbolize the elements to be clustered. The "next neighbour" relationship is expressed using arrows. This relationship is based on similarity functions [Backhaus 87] which are required for clustering and calculate the degree of similarity of two multi-dimensional profiles. In the context of the personal EPG, the function MATCH is used to determine the similarity of elements. For this purpose, MATCH is applied to pairs of elements. In a following step, the elements are grouped in such a way that centroids of neighbouring relationships are formed. The rectangles in Figure 29 illustrate the centre of gravity of the resulting clusters, i.e., the centroids (C_1, ... C_4). If two elements are their nearest neighbours, they are both connected to their second nearest neighbour. These computations are repeated until a minimum set of clusters has been achieved (depending on given threshold values). In our cluster example, the four sets (F,A,C,B,K), (G,I), (L,J,D,E,H,N), and (M,P,O,Q) are grouped. The resulting centroids (C_1, ... C_4) are used to adapt the user profile.

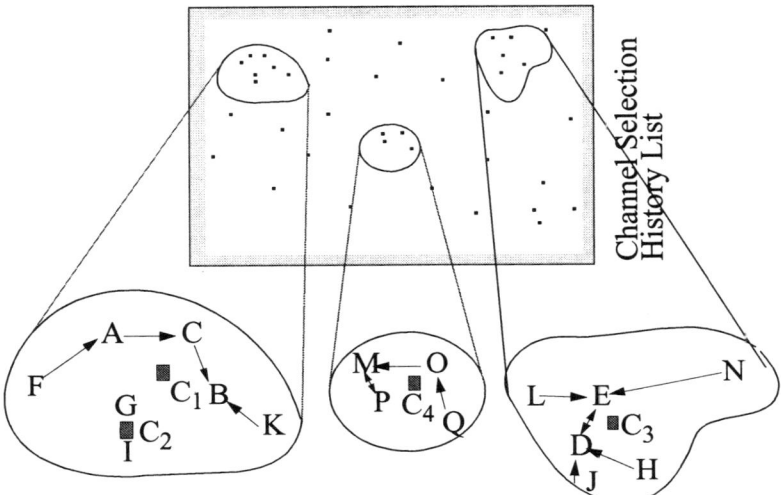

Figure 29: Clustering Method Used in Personal EPG

8 Profile Initialization and Update

8.1 User Profiles Initialization

User profiles are initialized when users subscribe to the Personal EPG service of the iMA. Based on the requirements R21.1 - R21.4 a range of modes for initializing user profiles are introduced.

8.1.1 Explicit Profile Specification

This mode is related to R16.1 and allows users to define their initial user profiles. The initialization procedure is shown in Figure 30. A form is given to the iMA user (1). This form is similar to internal structure of the iMA user profile database. The form is completed by specifying characteristic user requirements (2) and is directly stored in the user profile database of the iMA (3).

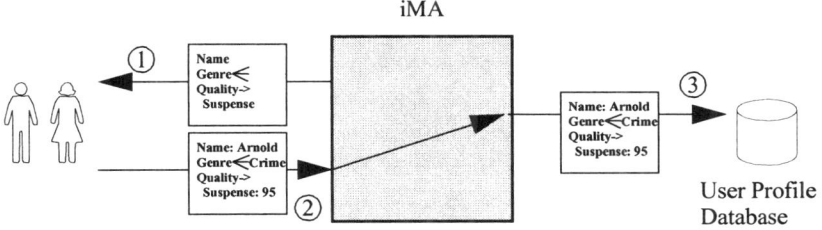

Figure 30: Explicit Profile Specification

8.1.2 Specification by Individual User Properties

Studies have shown that there is a relation between individual user properties (e.g., age, profession) and their interest in particular TV programs. By using specific user properties, initial user profiles can be generated. This mode fulfils requirement R16.2. As shown in Figure 31, users are requested to specify a set of individual properties (1). These properties are used by the iMA as input parameters for a look-up in a user property table ((2),(3)). The result of the table look-up operation is a characteristic user profile for the class of people to which the user belongs. Finally, the user profile is stored in the user profile database ((4),(5)) The structure, generation, and access methods of the Property/Program Table is shown in Section 8.2.

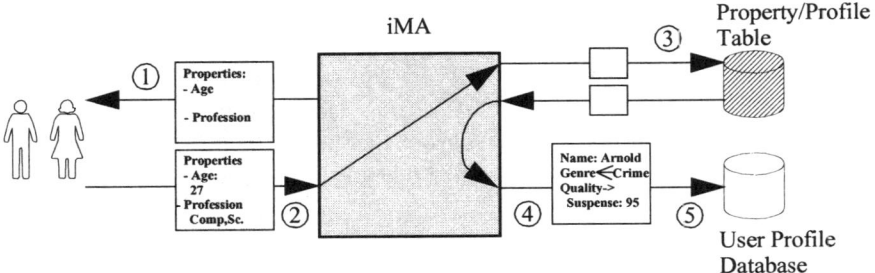

Figure 31: Initialization by Individual User Properties

8.1.3   Specification by Representative Program Examples

Examples of favourite programs are used to infer the personal TV program preferences (also see requirement R16.3). Figure 32 visualizes the process of initialization by using representative examples.

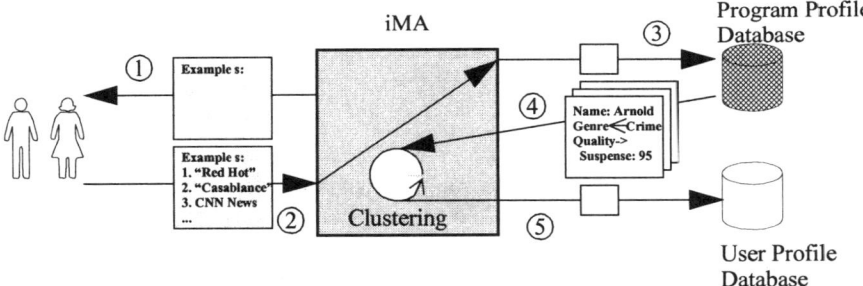

Figure 32: Initialization using Representative Examples

The user defines a list of TV programs which contains representative examples of his favourite programs ((1),(2)). The program profile of the examples ((3),(4)) are clustered. The resulting clusters are stored as user profiles (5).

8.1.4 Initialization by Observation

Based on the usage data, preferences and demands of the user are recognized. This mode allows an automatic initialization of the initial user profiles as depicted in Figure 33. In the observation phase the TV channel selections (i.e., time/date, channel identifier) are stored in a history list ((1),(2)). As soon as a sufficient number of user interactions have been stored the second phase begins. By taking the parameters in the channel selection history list as keys, ((3),(4)) the program profiles are read from the program profile database. The subsequent clustering derives the initial user profiles from these program profiles ((5),(6)). The amount of time a user watched a program is directly determines the priority of this program.

Phase 1: Observation

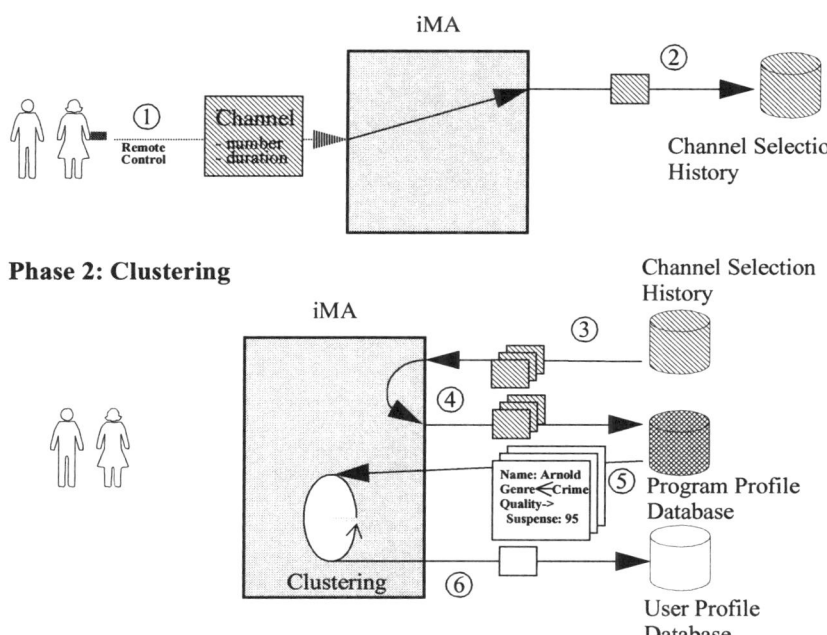

Figure 33: Installation by Observation

8.1.5   Comparison

So far, four initialization modes have been presented. The following evaluation criteria are applied (also see requirements R16.5 - R16.7):

- User interaction:
  How many user interactions in addition to the normal channel selection are necessary to set up the initial user profile?
- Complexity of calculation:
  What is the effort to calculate the initial user profile?
- Availability of results
  Time of availability of the initial user profile.
- Precision of results
  How precise is the initial user profile?

Table 6 shows the assessment of the user profile initialization modes.

Table 6: Comparison of User Profile Initialization Modes[a]

Initialization Mode	User Interaction	Complexity of Calculation	Availability of Results	Precision of Result
(1) Explicit Specification	⇑	0	immediately	⇑
(2) User Property-based	⇓	⇓	immediately	⇓
(3) Example-based	⇒	⇑	immediately	⇒
(4) Observation-based	0	⇑	later	⇑

a. ⇑.. high; ⇒..medium; ⇓..low

User Interaction

User interactivity is highest in mode (1), in which the users must manually specify every element of their user profiles. In mode (2), the degree of user interaction is reduced to minimum, whereas in the observation-based mode (4), no additional user interactions are required.

Complexity of Calculation

Mode (1) does not evoke iMA calculation efforts. The user profile attributes are calculatons taken from the manual specifications and are stored as the initial user profile. In mode (2), previously collected and stored property/profile mapping information is statistically analysed. There is minimal iMA calculation effort in the initialization phase (i.e., one table look-up operation). In modes (3) and (4) clustering processes are applied which require complex calculations. The complexity increases with the number of examples (mode (3)) and programs observed (mode (4)).

Availability of Results

The initial user profiles are immediately available in mode (1)-(3). The availability of the initial user profiles in mode 4 depends on the duration of the observation phase.

Precision of Results

The precision of initialization results in mode (1) and (4) is high. However, in mode (1), the precision depends on how well the user specifies each profile element.

8.1.6   Conclusion

There is no initialization procedure which completely fulfils requirements R16.5 - R16.7. The following recommendations are given for the application of initialization modes:

- Explicit Specification:
  Recommended for experienced users are already familiar with the organization of user profiles and are aware of the fact that creating a precise definition is time-consuming.
- Specification based on User Properties:
  Recommended for users unfamiliar with the iMA and user profile generation.
- Example-based Specification:
  Recommended for users wanting to create initial profiles of acceptable quality quickly.
- Observation-based Specification:
  Recommended for users unfamiliar with the iMA and user profiles, and those "technology-resistant" users who refuse to interact with the iMA. They are willing to wait for a certain period for their initial user profile.

The observation-based mode can be combined with other modes.

8.2   User Profile Reference Table

The user profile reference table is used to retrieve an individual user profile by means of specific user properties as input parameters. It is necessary to determine and define this set of user properties.[9]

8.2.1   Definition of User Properties and Attributes

Using statistical reports ([Feier 96], [Gerhards 96], [Grajczyk 96], [Media 94b], and [Oehm 96]) the following parameters have been shown to be the most characteristic properties of TV users: age, profession, education, sexuality, hobbies, and average time spent watching television.

Based on the significance and the practical value (similar to [Media 94b]) a subset has been chosen which is comprised of age, profession and educational level of users. This subset was used to define the user profile reference table of the intelligent media agent.

Age
Based on [Media 94a] the following age groups have been defined: 6-13, 14-19, 20-29, 30-39, 40-49, 50-64, over 64 years of age.

Type of Employment
The following cluster of professions have been defined:
- Group 1: manager, director, functionary, officer, executive
- Group 2: employee, servant, other profession
- Group 3: skilled worker
- Group 4: unskilled worker
- Group 5: no profession

Education
The following types of education have been categorized: (1) schooling without vocational training, (2) schooling and vocational training, (3) secondary schooling, and (4) university, college attendance, or technical schooling. A graphical representation of the user profile reference table is shown in Figure 34.

---

9. The definition and calculation of the user profile reference table was part of the diploma thesis [Göllner 95] which has been supervised by the author.

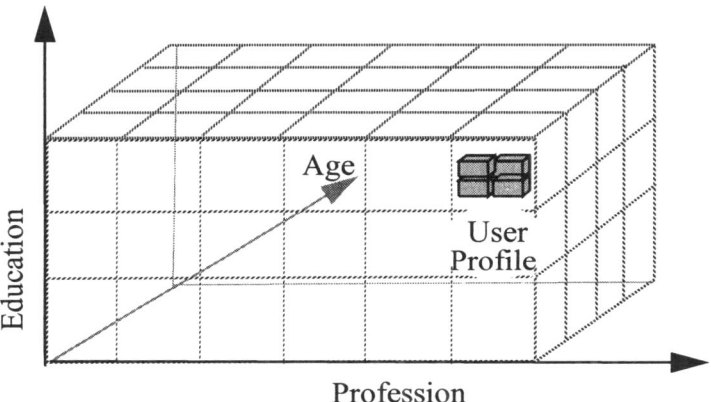

Figure 34: User Profile Reference Table

8.2.2 Definition of Initial User Profile Attributes

The user profile reference table has been defined and calculated within [Göllner 95] by using a program rating taken from the German rating database of Media Control Corporation [Media 94a]. This program rating is based on the hit list of all programs broadcast over the German TV channels ARD, ZDF, RTL, and SAT.1 in 1994.

An example hit list is given in Appendix D. All programs in the hit list have been described by using the data definition given in Section 6.2.

8.2.3 Results

Using the genre as an example user profile attribute and the age and type of employment as reference parameters, the user profile reference table was defined and implemented (for details see [Göllner 95]). The data and table definition was implemented using facts and rules of the decision support system CLIPS.

The user profile reference table can be copied or moved to other CLIPS run-time environments which satisfy requirement R27.

It is shown in [Göllner 95] that the functional requirements (as given in R16.2) are met. In the current implementation the user profile can be instantiated by putting the user's age and the type of employment into the set-top box.

8.2.4   Exceptions

Two examples taken from [Media 94a] are listed in Table 7 and 8 demonstrate the need for exception handling:

Table 7: Program Example A (ARD, 17.06.1994, Ages: 14-19)

Time	Name	Number of Viewers (in Mio.)
8.59-10.57pm	Soccer World Championship "Germany - Bolivia"	0.65
10.58 - 11.07pm	Soccer World Championship Game Summary	0.63

Table 8: Program Example B (ARD, 12.03.1994, Ages: 6-13)

Time	Name	Number of Viewers (in Mio.)
4.07-5.29pm	Disney Club	0.81
5.29-5.30pm	Spot - PR Campaign	0.63

Example A shows almost the same number of people were interested in the brief summary of the soccer game directly following the game's broadcast. In example B it is assumed that the high number of viewers of the "Disney Club" is the main reason of the relatively high number of PR campaign viewers (see also [Opasch 94]). The following conclusions can be drawn: Clips, commercials and previews are not regarded in the user profile reference table. Summaries, interviews and similar programs are considered of their own and regarded in the calculation of the user profile reference table.

8.2.5 Further Improvements

Precision

The elementary data [Media 94a] contains the top 20 programs per TV channel and category. As stated in [Göllner 95], this is not sufficient to achieve reliable results. Additionally, the calculation did not regard that seasonal events had influenced that program quotas (e.g., Olympic Games and Soccer World Championship). It is strongly recommended that any further extensions revise the user program reference table to take into account additional programs. Additionally, rules should be developed to detect and eliminate periodic events. [Media 94a] contains the hit lists related to the age, profession and educational level parameters. Instead of implementing a three-dimensional user profile reference table as depicted in Figure 34, it was necessary to implement autonomous user profile tables for each parameter. As a result, it is only possible to recognize specific interests based on one parameter of the user profile reference table. In any further extensions of the system, the hit list should account for all attributes of a user group identified in the user profile reference table.

Duration of Programs

The hit list given in [Media 94a] has listed programs independent of their duration. The duration of programs has not been regarded in the calculation of the user profile reference table. As result, an inversion of user priorities was expected to occur. The duration of a given program should be taken into account in subsequent extensions.

Additional Parameters

So far, several user parameter have not been taken into account. In order to achieve more precision additional parameters should be introduced. Statistical reports ([Feier 96], [Gerhards 96], [Grajczyk 96], [Media 94b], and [Oehm 96]) have shown that sexuality and hobbies are additional significant parameters.

8.3 User Profile Updates

The update modes are similar to the user profile initialization modes described in Section 8.1. An additional parameter is of specific interest when updating a user profile, namely the time period or window during which the user selections are of interest. For the update of user profiles the following modes are distinguished:

(1) Every Program Mode: All programs a user has viewed are stored in the history list. This approach is very inefficient in terms of memory and computation time requirements. Old and new programs are used in the adaptation of new user profile. New programs can only slowly be regarded in the user profile due to dominant number of old programs.

(2) Wrapping Window Mode: Each program viewed since a certain point in time (e.g., the last update) is regarded in the adaptation process. Depending on the window's size, old and new programs can evenly be mixed. However, the program information is removed as soon as the wrapping window expires. Therefore, the precision depends on the actual time and the window size. Hence, the availability of the wrapping windows method is lower than in case of mode (1) and (3). Programs which were broadcast previous to start time of the window are not taken into account. The wrapping window method can also be applied to the number of programs seen.

(3) Sliding Window Mode: The last n programs seen or the programs viewed in the last n time slots are used for the profile adaptation. Depending on the window's size, old and new programs can evenly be mixed. The resource requirements are low. However, programs which were broadcast before the start time of the window are not taken into account.

Comparison

The criteria for comparing the three modes with one another assess the resource requirements, whether old and new programs are considered in the update, and the mode's availability as a measure of its convenience. Availability specifies to what degree a spontaneous execution generates accurate results.

Table 9: Evaluation

Criteria	Every Program	Wrapping Window	Sliding Window
Resource Requirements	⇑	⇓	⇓
Consideration of Old Programs	⇑	⇓	⇓
Consideration of New Programs	⇓	⇑	⇑
Availability	⇑	⇒	⇑

Because of its low resource requirements, good consideration of new programs, and high availability, the sliding window was chosen as windowing mechanisms in the automatic update process of the iMA. For special programming events whose period is longer than that defined for the window being used (e.g., World Soccer Championships, Olympic Games). an exception handling maintains a separate program history list.

Similar to the initialization modes, user profiles can be updated by (1) updating the user profile attributes, (2) updating the individual user properties, (3) updating the list of representative television programs, or (4) automatically clustering the interests derived from the history list.

8.4   User Groups

Users can share their user profiles and form user groups (see requirement R19). There are explicit and implicit user groups.

Explicit User Groups

An explicit user group is set-up by users who already have individual user profiles and want to merge them. For example, two friends are interested in finding a movie both would be entertained. In the iMA, explicit user groups are organized by clustering the user profiles of all group's members.

Implicit User Groups

If no additional identification is required for group members (e.g., there is only one identifier associated with the whole family) the user identifier must necessarily be shared by all members of this group. To iMA, this group appears as one user.

9 Decision Support System

The model presented in Section 7 is predicated on the assumption that the longer a program is viewed, the more likely that this is or will become a viewer's favourite. It has been found out that there are exceptions to this assumption. This section analyses these exceptions and shows how they can be handled by the intelligent media agent. Extensions to the matching and clustering functions are proposed.[10]

9.1 Characteristic Properties

So far, the automatic initialization and update modes have been described in Section 8. In both modes, clustering is used to derive user profiles from the programs seen. The duration a user watched this program was assumed to be proportional indicator of his interest. In the following, exceptions are given where this assumption is not applicable.

9.1.1 Visitors

The specific situation of viewing done by visitors was described in the grandson/grandfather example in Section 3.3 on page 33. The term is synonymously used for all other people using the television set of the user or user group for whom the user profile has been generated.

Because visits can substantially influence the watching behavior, changes in the usage statistic can be expected, i.e.: (1) The properties of the program watched (e.g., time, genre) different from the existing user profile. (2) The behavior of the user related to the parameters of the interaction process may differ.

9.1.2 Program Changes

If the program announced is different from the program sent (e.g., due to a last-minute program change), the program database is suddenly rendered inaccurate. Within the intelligent media agent it is therefore necessary to receive program updates as soon a programming change is announced. This is indicated by a notification message containing the new program descriptor for the program changed.

---

10. This section is based on the diploma thesis [Vazirian 95] which was supervised by the author.

9.1.3   Periodic Events

Due to the dependency of user interests from the time of day, day of week and holiday, these parameters have already been included in user profiles. In addition to these preferences, it is also possible that the interest is related to (1) specific programs which (2) have a long period. For example, the Soccer World Championships is extremely popular event recur every 4 years (see [Göllner 95]).

Example descriptors of periodically recurring events are given in Table 10. Every time a personal television program is calculated, the intelligent media agent must check whether a periodically recurring program is a favourite program.

Table 10: Descriptor of Periodic Programs

Name	Start Time	Duration	Period
Soccer World Championship	1994	2 weeks	4 years
Olympic SummerGames	1996	3 weeks	4 years

9.1.4   Zapping

Zapping is the name of the process given to a use's rapid switching between television channels. When considering zapping as an exception, it is necessary to analyse the reason behind zapping. [Krüger 96a] describes the following trends in the characteristics of television programs:

- more fragmented programming
- more scenes and cuts in programs
- a higher percentage of advertisements, trailer, header

For example, the average number of program units per hour have been increased from 5,7 in 1988 to 13.2 in 1995. Figure 35 [Krüger 96a] compares two schedules of television programs from 1985 and 1995. Changes in the program are marked with black bars.

Results of psychological research related to zapping motives can be found in [Jäckel 93] and [Vorderer 94]. For example, telemetrical measurement test cases (see [Vorderer 94]) have shown that channel zapping depends on the emotional involvement in the program: Channel switching occurred after a low emotional involvement of the test persons in the television program for 2 to 4 minutes. This result is only applicable if the emotional involvement has already passed a minimum

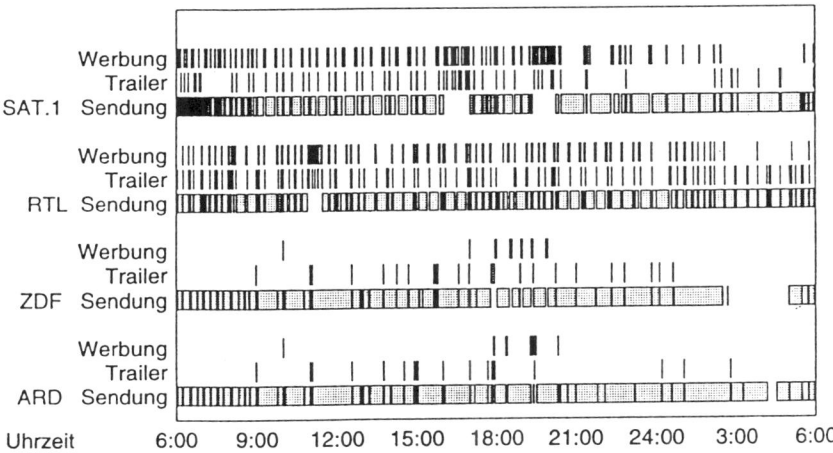

Figure 35: Program Schedule of ARD, ZDF, RTL and SAT.1

threshold. Similar dependencies have not been found for the cognitive involvement in the program. The expectation of positive involvement in the program does not necessarily lead to a continuous viewing of that program.

The following motives for zapping can be summarized by:

(1) Preview of programs: As depicted in Figure 36, the user watches the program on each channel for a certain time in order to preview all current programs. Determining which actual program is of the most interest is the main objective of zapping. If a program of interest is recognized as the favourite program of the moment, this program is then selected and watched.

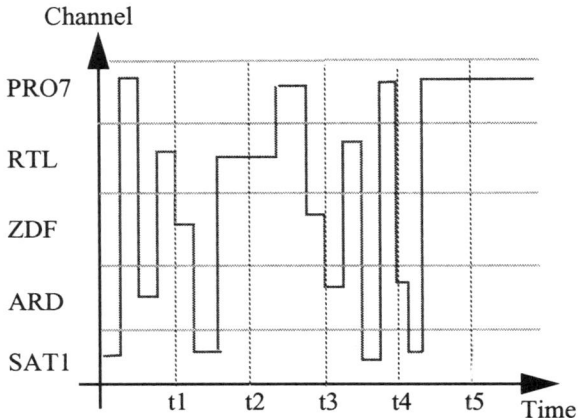

Figure 36: Program Preview Example

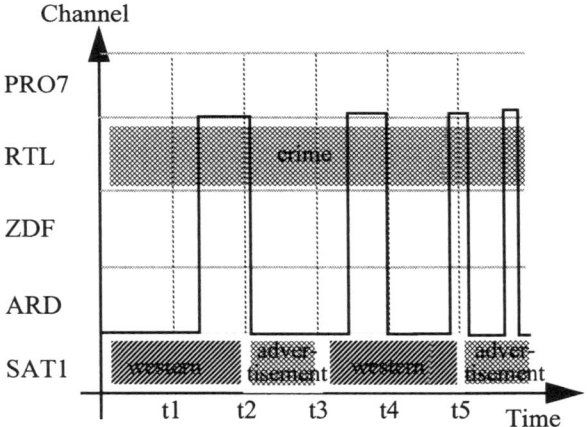

Figure 37: Watching Programs in Parallel

(2) Watching programs in parallel: Assuming a TV device has no picture-in-picture function, channel zapping can be used to watch a set of television programs in parallel. A typical example is given in Figure 37. As indicated below, there are frequent switches between the two favourite programs on SAT.1 and RTL. The average duration of watching is higher than in the preview mode (see (1)).

(3) Interruptions: As part of the program broadcasting schedule, frequent interruptions (e.g., commercials, clips) may occur. Statistical measurements (see [Opasch 94]) have shown that 80 percent of the people are not watching television (e.g., leave the room, switch the TV channel) when commercials are shown. The example of channel switching is given in Figure 38. Interrupts in the program are used to zap through the channels. This motivation is different from (1) where the user intents to search for better programs. Here, the only goal is to avoid watching advertisements. As soon as the interrupt is finished and the favourite program continues, the user returns to the program.

As a characteristic property of this mode, the user typically returns to the favourite channel after a certain period of time. It has been experienced that the frequency of the switches returning to the favourite channel increases if the expected average duration of interruption has nearly been reached.

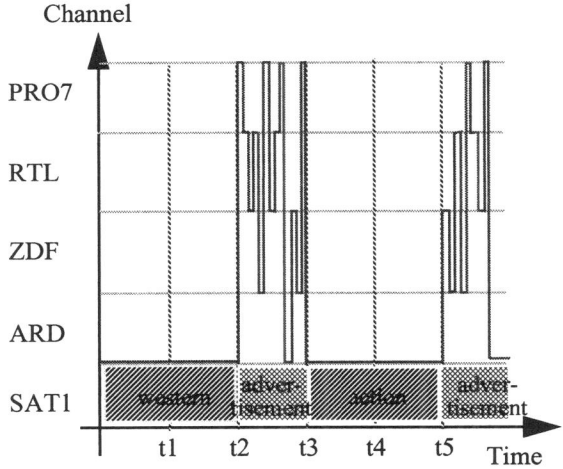

Figure 38: Interruptions Example

9.1.5   Sleeping

Suppose that a user is sleeping while a TV program is turned on. A sleeping user is assumed to can be recognized if there is no user interaction for a long time though his favourite programs are appearing on other channels. Additionally, the probability of a sleeping user is higher in the evening hours (see also Figure 39).

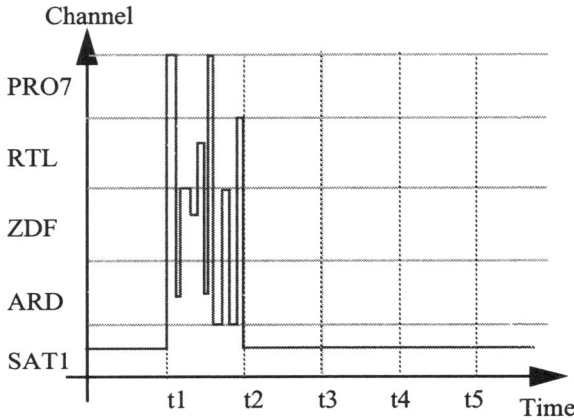

Figure 39: Sleeping Example

9.2   User Behavior Variables

It is necessary to define a user interaction model to establish any decision support system. In the following it is shown how the individual behavior of users can be incorporated in variables.

Zapping Model

Based on the analysis of zapping in the previous section, there are two phases which should be considered: bursts of zapping operations and phases with minimum or no zapping. The following parameters can be defined:

Minimum Switch Interval (MISI):

Minimum time between two switches. If the time between two subsequent switches is below MISI, the channel switch is ignored.

Maximum Switch Interval (MASI):

Maximum time allowed between two switches in the same burst phase. If the time between two subsequent switches is greater than the MASI than this burst phase has ended.

The zapping model (see Figure 40) defines the specific variables:

Burst Duration (BD):
Duration of a burst period.

Burst Gap (BG):
Time between two subsequent bursts.

Number of Switches (NS):
Number of switches during a burst.

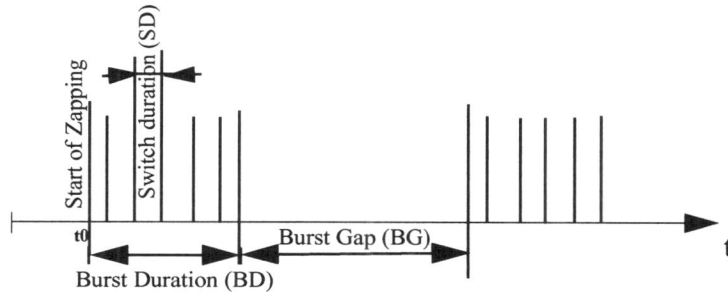

Figure 40: Zapping Model

The averages of the parameters BD, BP and NS can be calculated and used to derive additional parameters:

Average Program Watching Time (APWT):
Average time a program is watched. This parameter can be applied to the burst duration and the burst gap.

Average Burst Period (ABP):
Average period between subsequent bursts.

9.3 Reaction to Zapping

Predicated on both a user behavior model and the variables, a rule-based system is introduced to recognize zapping and to consequently initiate a set of actions.

Examples of zapping detection and reaction rules are given in Table 11. The first rule in the table describes how channel zapping is detected. If the conditions are true, then the iMA has detected the beginning of a burst phase.

The second rule registers the program descriptor for the profile update operation as soon as a specific program has been seen for a long time (APWT = high). Depending on the update mode, the profile of this program is registered to be added or merged with existing profiles.

Table 11: Examples for Burst Detection and Reaction

Condition	Action List
(event = channel_switch) (time_since_last_switch > MISI) (time_since_last_switch < MASI)	(burst = ON)
(event = channel_switch) (time_since_last_switch > MISI) (time_since_last_switch > MASI) (burst == OFF) (APWT == high)	(register_profile.channel_desc)

10 Implementation and Testing

Two versions of multimedia interfaces have shown that the concept of a personalized multimedia application is applicable and usable based on both the MHEG standard and the HTML/JAVA approaches.[11]

10.1 Implementation based on the MHEG Standard

A version of the personal electronic program guide has been implemented within the "Globally Accessible Services (GLASS)" project. GLASS is an interactive multimedia and television system which was one of the first systems worldwide to be based on the MHEG standard (details see [Berkom 94], [Cossmann 96]). In addition to the entity which computes the MHEG presentations (also called "MHEG engine"), various complementary functional modules have been designed and implemented.

These modules can be subdivided into three categories:
- Client Components, consisting of the User Interface Agent (UIA), Presentation Objects (POs), the MHEG engine (engine) and the Control Agent (CA);
- Management Components, consisting of the Session Management Agent (SMA) which, in turn, includes the Authoring Agent, the Security Agent, the Locator Agent, the Directory Agent, and the Data Distribution Agent; and
- Stores consisting of MHEG object stores and Content Data Stores.

The intelligent media agent includes the Virtual Object Store and is responsible for dynamically generating MHEG objects which implement the personal electronic program guide application. Figure 42 gives an overview of the system architecture including the intelligent agent functions:

The User Interface Agent is responsible for the management of Presentation Objects. Presentation Objects represent specific multimedia contents, e.g., the audio, video, animation, text, and interactive elements. The MHEG engine is a central component. It is responsible for loading and interpreting multimedia applications and for initiating corresponding actions and sending them to the User Interface Agent.

A so-called Control Agent is part of the MHEG engine managing all communication aspects. The MHEG objects are received from the Session Management Agent. The Session Management Agent contains sub-components which are responsible for

---

11. Parts of this section have been published in [Cossmann 96].

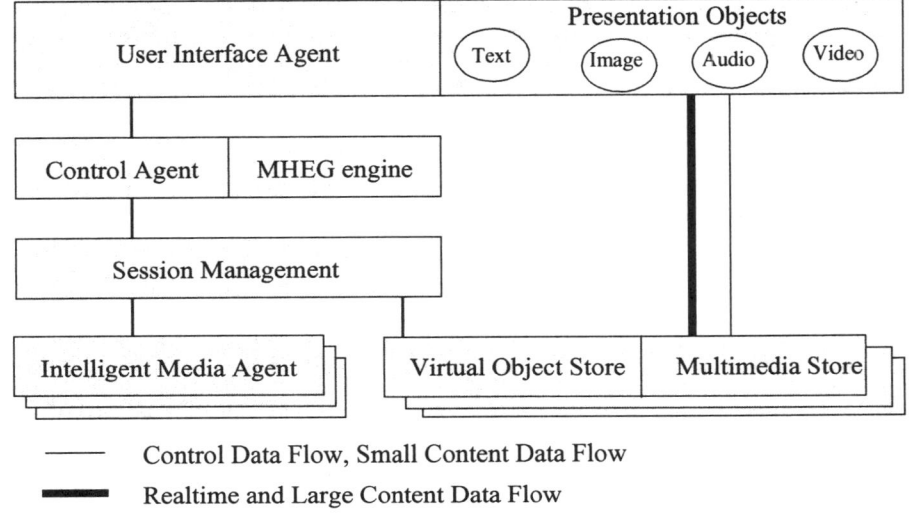

Control Data Flow, Small Content Data Flow

Realtime and Large Content Data Flow

Figure 41: Architecture of the MHEG-based System

solving the problem of data distribution, security and accounting. The entity which is responsible for content storage and on-demand-delivery of discrete or continuous media is called Multimedia Store. The following sections will briefly introduce the functionality of these components. Further details can be found in [Cossmann 96].

10.1.1   User Interface Agent and Presentation Objects

The User Interface Agent (UIA) is a component residing at the client site (e.g., the set-top box of interactive television systems). The User Interface Agent's most important tasks include the creation, maintenance and destruction of Presentation Objects. The initiation of these operations is initiated by the MHEG engine. The User Interface Agent and the MHEG engine are interconnected by using communication protocols which are based on standardized communication protocols (e.g., UICP - User Interface Control Protocol which is an ASN.1-specified presentation protocol, transmission control protocol, Internet protocol). For example, the UICP protocol is comprised of primitives for session establishment, control and accounting, as well as primitives for Presentation Object control. A typical transaction consists of a request protocol data unit that is asynchronously answered with a response protocol data unit. Requests can be originated by the MHEG engine which instructs the UIA to perform certain presentation tasks. Most of the UICP PDU's represent MHEG actions (see [ISO 135225]). PDU's can also be generated by the UIA. They can be used to capture and transfer an event (e.g., mouse moved, keyboard pressed) or to send data to the SMA.

Presentation Objects (POs) are typically responsible for presenting content data objects and the any user interaction with these objects. PO's are created, modified and destroyed on demand by the UIA. Content data objects, managed by Presentation Objects, can be extracted from an MHEG object, if included, or retrieved from the Content Store, if referenced in the MHEG object. In the latter case, Presentation Objects communicate with the Multimedia Store by using the Presentation Object Control Protocol (POCP) and the Presentation Object Data Protocol (PODP). The POCP establishes and controls the PODP which performs the actual data transmission. While POCP primitives such as "open data connection", "start/stop data streaming", "set stream speed" can be compared with the control functions of a VCR-type device, the PODP transmits digital media data. In the context of GLASS, the PO's were implemented to present content and interact with the following media types:

- Video (MPEG I),
- Audio (MPEG-Audio and WAVE),
- Images (JPEG), and
- Text (plain text and as well as the GLASS specific text format)

PO's can be classified into two subsets according to media type — discrete or continuous. Video and audio are examples of continuous media types. Due to resource constraints in the client systems, they may neither be cached nor instantiated more than once. Examples of discrete media that can be preloaded when the PO is instantiated are images and texts.

10.1.2   MHEG Engine

The MHEG engine is an important component which reads and loads MHEG applications. To execute this task, the MHEG engine interprets the interchanged MHEG objects. As a result, it issues requests to the presentation system. The interpretation process can result in requests for retrieval of further objects.

The Engine requests MHEG objects asynchronously. The Control Agent (CA) is responsible to delegate these requests to the MHEG object stores and the Session Management Agent for retrieving these objects. The result of this retrieval process is an asynchronous response as an input to the MHEG engine. As described above, events are captured by the User Interface Agent. These events are passed to the MHEG engine. MHEG objects and events are processed by the Engine according to the actions defined by the MHEG application. For example, this results in a request to the presentation systems or starts the retrieval of additional MHEG objects which describe the next parts of the multimedia application. From an internal point of view, the

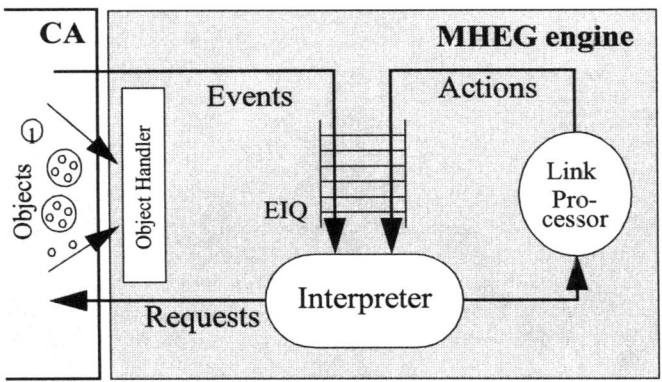

Figure 42: Architecture of the MHEG engine

MHEG engine manages an event queue which serializes all incoming and internal events received from the presentation system, all notifications of MHEG objects provided by the CA, and all actions that are triggered by internal state changes which are the result of processing such incoming information.

10.1.3   Multimedia Store

The Multimedia Store is responsible for the storage of content data and the transmission of this at the request of its clients. The Multimedia Store provides the following functionalities:

- control of content data delivery to the Presentation Objects, and
- transfer of continuous-media data.

The communication with Multimedia Stores is done using PDU's. The Multimedia Store consists of two components:

- Video-on-Demand (VoD) server including appropriate stream handlers to manage of multimedia data, i.e., the storage and transmission of continuous-media data, and

- Control Process which communicates via specific protocols with other system components (i.e., POs and SMA) and instructs the VoD server accordingly to deliver the required multimedia data stream.

The Multimedia Store can ensure the delivery of continuous media streams. A resource management system is used to make certain that the quality of service requirements are met (see [Wolf 95]). The Multimedia Store is predicted on a file system with resource management capabilities which enables the integrated management of continuous media requirements and provides generic resource control facilities.

10.1.4   Management Components

The Session Management components are responsible for organizing such management tasks as directory services, data distribution services, or the retrieval of MHEG and content data objects. Object delivery from the Multimedia Stores via the Session Management and Control Agent to the MHEG engine has already been described above. The main method to transmit large objects (e.g., video files, images) is as follows: using uniform resource locators objects are referred to the file in the Multimedia Store which contains the media data. Data transfer is done directly between the Multimedia Stores and the Presentation Objects using the PODP protocol.

The Store Control Protocol (SCP) specifies the static elements of the communication between the Session Management components and Multimedia Stores. This protocol contains primitives for the initiation of data transfers between Presentation Objects and Stores/Gateways, for the retrieval of MHEG objects from MHEG Stores, and for accounting purposes.

10.2   Implementation of the Electronic Program Guide Application

The personal electronic program guide application of the intelligent media agent has been implemented as an MHEG- and as an HTML/JAVA-based prototype system. The system and user behavior have been investigated by using both prototype implementations. While the MHEG version has been tested in a local testbed, the HTML/JAVA-based version was available for a broad community of Intranet test users. The implementation and tests were based on fixed program descriptors.

Switching the television programs was emulated using a video-on-demand and a text mode, i.e., as soon as a program was selected by the test user, a video file containing a program preview was started. A textual preview was given if no preview video was found in the database.

The intelligent media agent was implemented using the script language Tcl/Tk and the relational database POSTGRES. The decision support system is based on AgentClips - an agent extension of the decision support system, CLIPS being chosen as basic agent operating system[12].

The prototype implementation of the intelligent media agent can be broken down into the following components:

1. A metadata base containing attribute names, types and parameter descriptions of the program descriptor and user profile. Add, change or delete operations are implemented.
2. This metadata base is used by an administrative tool to generate the program description and user profile database.
3. The functions MATCH and CLUSTER supply the basis for the initialization and update modes.
4. The user profile reference table was implemented using the rule-based AgentClips environment.
5. An administrative tool was implemented based on the Tcl/Tk scripting language and on HTML/JAVA (see Appendix C), thereby granting the administrator access to program descriptors, test user profiles and calculation parameters in the test environment and specify parameters of the clustering process.
6. Approximately 3,000 movies classified by genre, 100 movies classified by quality properties, as well as 10 test user profiles were used.

10.2.1   MHEG Prototype Implementation

Figure 43 shows the user interface implemented utilizing the MHEG standard. A hit list is displayed which lists the top oay-per-view movies for the individual user. This hit list has previously been calculated based on the personal user profile containing the genres crime and adventure as the preferred genres. On the right-hand side of the figure, two types of navigational elements are offered: The buttons "Menu" and

---

12. POSTGRES was developed at Berkeley University, California, CLIPS by the US NASA, and AIX is trademark of IBM.

"Back" can be used to leave the personal electronic program guide application. The buttons "All Channels", "Movie Channel", and "Pay View" are selector buttons which can be applied to the hot-list of movies.

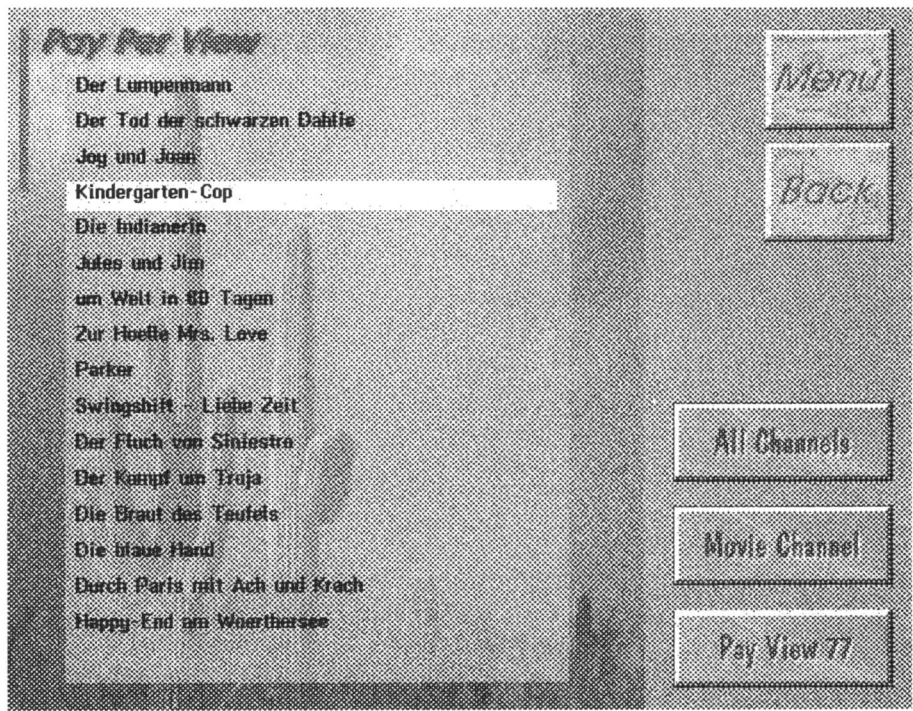

Figure 43: Personal EPG Prototype Utilizing the MHEG Standard

10.2.2   HTML/JAVA Prototype Implementation

A sample hot-list showing the user interface of the HTML/JAVA-based implementation is depicted in Figure 44. This page is based on the previous selection of a personal user profile containing the parameters time (24.6.1996, 10-12am) and the program preferences (news or series) which were applied to the German television broadcast channels. For example, the program "Ein Colt für alle Fälle" is part of the personal hit list because it is a series and because it is broadcast during the time specified in the user profile. Additional examples are shown in Appendix B.

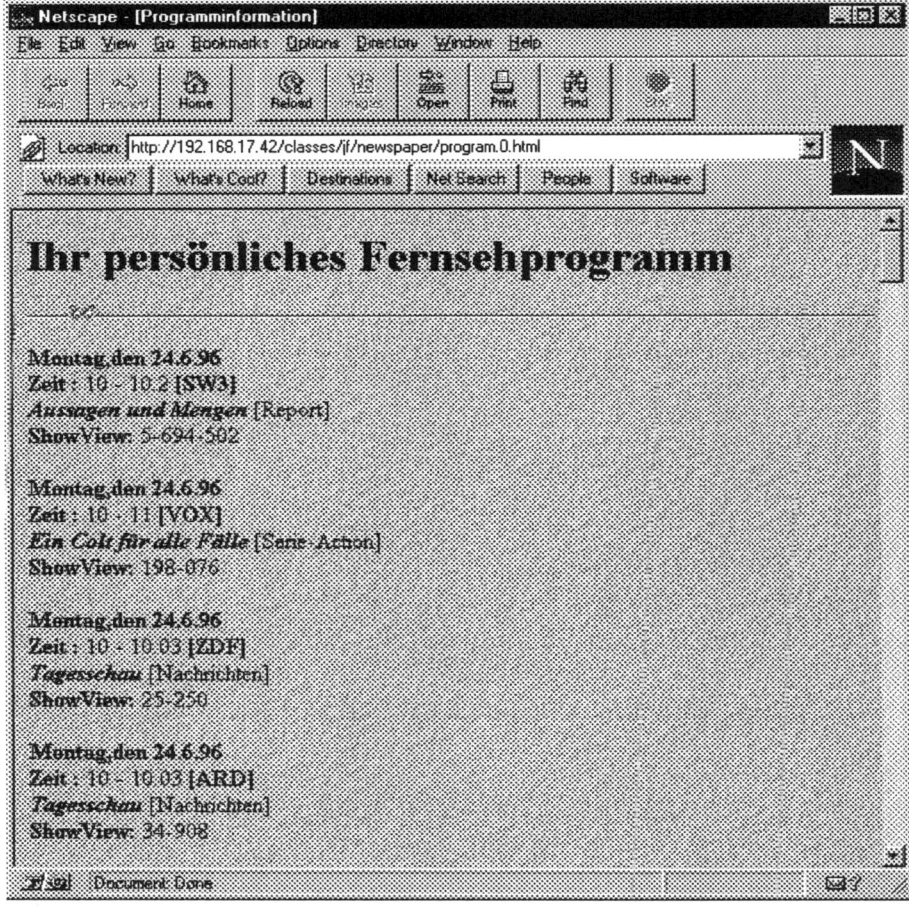

Figure 44: Personal EPG Prototype Based on the JAVA Standard

10.2.3   Comparison of MHEG and JAVA Approach

A comparison of MHEG and JAVA was given in [Dahm 96]. Additionally, [DAVIC 95d] describes a proposal combining both MHEG and JAVA in an interactive television framework.

In this context, the implementation of both versions has shown that both approaches are sufficient to implement a personal electronic program guide. Table 12 shows an evaluation applying portability and functionality (user interface, intelligent agent, management) criteria.

Table 12: Comparison of Presentation Modes

Criteria	MHEG	HTML/JAVA
Portability of Application	⇑	⇑
Functionality of User Interface	⇑	⇓
Intelligent Agent Functions	⇓	⇑
Support for Management Functions	⇓	⇑

Both versions are based on a portable application format which can run on multiple heterogeneous platforms. The MHEG-based environment provided for the rich functionalities of multimedia application programming, while the HTML/JAVA-based environment does not possess the control components necessary for rich multimedia application programming. One consequence of this is that there is no integrated JAVA-HTML-based solution of presenting video files. However, with the implementation of intelligent agent functions, it is JAVA which allows for the programming of complex tasks in addition to the provision of a multimedia run-time environment. In an MHEG-based approach, the MHEG run-time environment must be extended by an MHEG-3 compliant script processing engine. Therefore, the management functions can not easily be implemented at the client site in the MHEG-based mode while the use of HTML/JAVA allows for programming at the client and server sites.

10.2.4   Results of Clustering[13]

Specific attention was given to the tests of the automatic installation and update modes as well as the clustering method. The following test scenario was applied: Test users reported the TV programs they watched for three weeks. The test report was then sent to the administrator of the personal EPG application and inserted into the user history list of the system by using the administration tool (see Appendix C). Applying the clustering method to the history list, the resulting cluster hierarchy was displayed as and as histogram.

The test cases are based on the following parameters: Genre, time of day, and duration of watching have been used as clustering criteria; Ward's clustering method was applied. The threshold for zapping detection was defined as 10 min. Events which were watched for less than 10 min were not considered in the clustering process. The threshold of the minimum distance between clusters was defined depending on the number of events in the history list. The program descriptors were defined using program metadata on the Internet. This data contains less information than that defined in Section 6.2 (e.g., genre data).

10.2.4.1   Test User A

The application of the clustering algorithm presented in Section 7.4 resulted in a hierarchy of clusters. Dendrograms were used to visualize the hierarchy of clusters. The dendrogram of test user A is shown in Figure 45. In the horizontal direction, each of the 45 events in the history list is depicted. The vertical axis represents the distance between clusters. In order to aggregate user profiles, it is necessary to specify when the clustering process should end. If the threshold is low then there are many clusters containing very specific interests. If the threshold is high, then there are just a few clusters containing broad interests. In the current implementation the threshold parameter can be dynamically optimize. In Figure 45, a threshold=2.25 was applied and resulted in 9 clusters[14].

For example, the most similar event to event number 26 is event number 22. Both events belong to cluster 8. In order to analyse and visualize the clusters in more detail a cluster monitor was implemented and used (see Appendix C).

---

13. A detailed presentation of test results in given in [Michel 97].

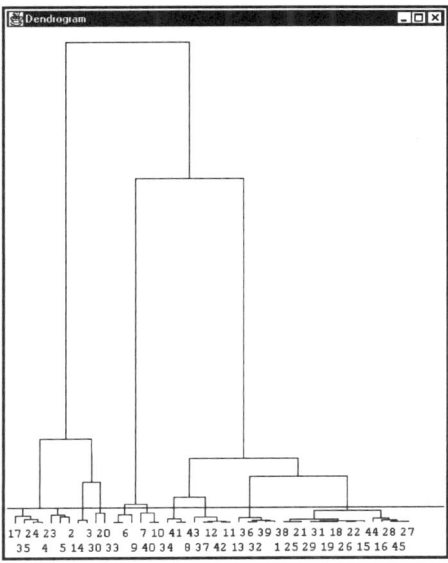

Figure 45: Dendrogram of Test Results of Person A

Figure 46 depicts the cluster monitor applied to the clusters of test user A. The top diagram shows all the clusters sorted by number. The y-axis specifies the number of elements in the cluster. For example, the most significant cluster for test user A is cluster 8 which mainly consists of evening programs around 8 pm with the genre "news" and an average duration of 14 minutes. The scroll bar on the bottom of the cluster monitor is used to display the details of the cluster and its events: cluster number, date and time of event, percentage of duration the user watched the event, and genre.

---

14. The standard parameters of the clustering process are defined as follows (details see Appendix C):
period = Mon-Sun
weight of time of day, genre, and duration = 1
automatic selection = on
clustering method: Ward's method

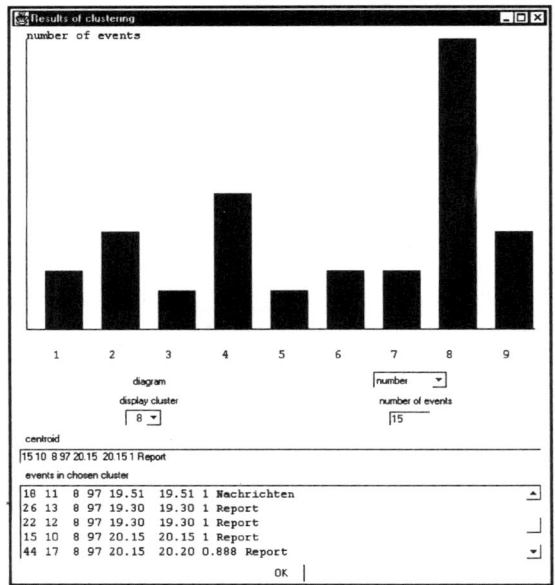

Figure 46: Cluster Monitor of Test Person A

In the example, it can be seen that events 26 and 22 are very similar. Both events describe the same program which was seen twice on the days (97/08/12 at 7.30 pm and 97/08/13 at 7.30 pm).

A detailed summary of clusters is given in Table 13. Comments are given to the clustering results. Beside cluster 8 the clusters 4, 2 and 9 are important. In addition to cluster 8, cluster 4 mainly contains report programs. Both clusters differ in the time of day (8.15 pm versus 9.15 pm) and in the duration of watching (100 versus 67). In terms of method, the clustering process produced results which reflect the accumulated interest of the test user. A need for improvement is found when analysing the genre "reports" in more detail. The test user was specifically interested in regional and travel reports. This information could not be regarded in the personal EPG application because only one level of genre information was attached to the program information by its provider. A system should be implemented which applies a fine granularity to the program information as identified and proposed in Appendix B.

Table 13: Cluster Table: Test User A[a]

Number	Centroid of Cluster			Comments
	Start Time	Duration % Watched	Genre	
1	1.20 pm	100	Report	
2	8.15 pm	100	Movie	Movies in the evening hours are important.
3	4.30 pm	57	Sports	
4	9.15 pm	67	Report	Separated from cluster 8 because of less interest and different starting time.
5	7.53 pm	17	Entertainment	Seen between two reports.
6	10.35 am	100	Report	
7	5.00 pm	100	News	
8	8.15 pm	100	Report	Main interest.
9	18.00 pm	100	Sports	Sports are important.

a. Parameters: The minimum distance between clusters is set at 2.486, the weights of the time of day, duration and genre are set to 1.

The clustering process can be configured by modifying the weights attached to the clustering parameters time of day, genre, and duration. For example, if the duration of watching is not regarded in the clustering process, the weight parameters are defined as follows: time of day = 1; genre = 1; duration = 0. The result of the clustering process is shown in Figure 47.

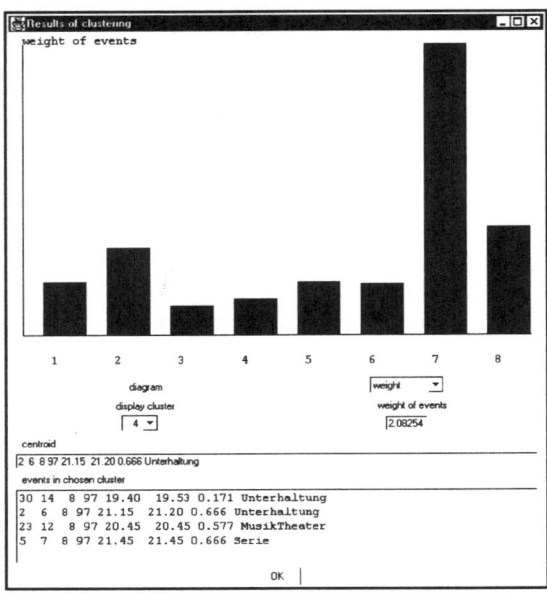

Figure 47: Cluster Monitor of Test Person A (duration = 0)

Details of clusters and their centroids are given in Table 14. Compared with the structure shown in Table 13, major changes occurred in clusters 4 and 8. Two clusters were calculated for sports programs in Table 13 (cluster number 3 and 9). When ignoring the duration of events, they are summarized in one "sports" cluster (cluster number 8). Further test cases (e.g., time of day = 0) are given and discussed in [Michel 97].

Table 14: Cluster Table: Test User A; duration = 0[a]

Number	Centroid of Cluster			Comments
	Start Time	Duration % Watched	Genre	
1	1.20 pm	100	Report	

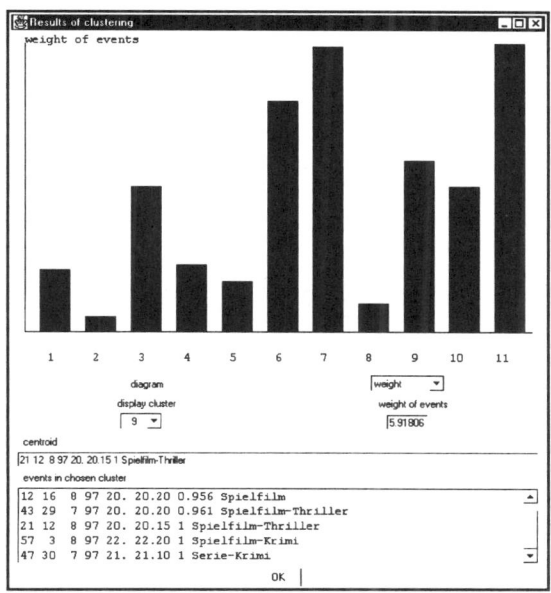

Figure 48: Cluster Monitor of Test Person B

Table 14: Cluster Table: Test User A; duration = 0[a]

Number	Centroid of Cluster			Comments
	Start Time	Duration % Watched	Genre	
2	8.25 pm	92	Movie	
3	11.00 pm	67	Report	
4	9.20 pm	67	Enter-tain-ment	Cluster consisting of genres which have been watched not often.
5	10.35 am	100	Report	

Table 14: Cluster Table: Test User A; duration = 0^a

Number	Centroid of Cluster			Comments
	Start Time	Duration % Watched	Genre	
6	5.00 pm	100	News	
7	8.15 pm	33	Report	
8	6.00 pm	100	Sport	Two sports clusters have been integrated into one cluster.

a. Parameters: The minimum distance between clusters is set to 2.38, the weights of the time of days, and genre are set to 1; duration = 0.

10.2.4.2 Test User B

The clustering results of test user B are shown in Figure 48. A detailed overview of the clusters is shown in Table 15. The favourite genre of test user B is science fiction. Comedy, thriller movies, and reports were also of interest to him. Evening is the preferred time of watching. Some favourite programs were watched in the afternoon (e.g., "Starship - The Next Generation"). If test user B is interested in the program, he watches this program completely.

Table 15: Cluster Table: Test User B^a

Number	Centroid of Cluster			Comments
	Start Time	Duration % Watched	Genre	
1	8.04 pm	73	News	
2	10.14 pm	32	Music	Several music programs
3	8.00 pm	67	Report	Evening News and Reports

Table 15: Cluster Table: Test User B[a]

Number	Centroid of Cluster			Comments
	Start Time	Duration % Watched	Genre	
4	8.25 pm	75	Movie -> Science Fiction	
5	3.30 pm	50	Movie -> Science Fiction	
6	6.25 pm	100	Movie -> Comedy	
7	3.00 pm	100	Movie -> Science Fiction	Program: "Starship - The Next Generation"
8	8.07 pm	53	News	
9	8.15 pm	100	Movie -> Thriller	Test user prefers movies in the evening (e.g., thriller, adventure, ..).
10	11.05 am	100	Movie -> Science Fiction	
11	8.15 pm	100	Movie -> Science Fiction	Program: "Akte X"

a. Parameters: The minimum distance between clusters is set at 2.498, the weights of the time of day, duration and genre are set to 1.

10.2.4.3 Summary

The clustering approach has been tested. Test scenarios demonstrated that clustering methods can generate a personal user profile based on the history list of programs. This profile can be used as input to the MATCH process to generate a Personal EPG by comparing the user profile with the program descriptors. It was shown how parameters are used to optimize the clustering process: The threshold of the minimum distance between clusters can be used to control the number of clusters generated. For example, it may be useful to define this number depending on the overall TV consumption of the individual user. The priority of the duration, genre, and time of day can be configured by attaching weight parameters. For example, if the preferred time of day is not as important as the other parameters for the Personal EPG, the weight assigned to duration can be reduced.

The test cases could also be applied to an extended set of parameters. If a fine granularity is provided in the program descriptors (especially in the genre parameter), then a more precise configuration is possible. It is also possible to regard the length of the program in addition to the duration parameter.

It has been shown that the clustering approach is applicable in the context of the personal EPG application. In addition to the results presented, [Michel 97] contains the results of more tests, and an evaluation of the resulting user profiles done by the test persons which indicate that their personal interests are taken into account in the user profiles generated by the iMA. In Section 4.6 it was pointed out that no clustering algorithm can generally be recommended, and that the specific clustering method should be selected based on the application scenario. The following algorithms have been implemented and tested in the clustering module of the iMA:

- single-linkage-method,
- complete-linkage-method,
- average-linkage-method,
- weighted-average-linkage-method,
- median-method,
- centroid-linkage-method,
- Ward's method, and
- Lance/Williams-method.

By using the examples given in [Henrion 88] the clustering methods have been tested. The application of Ward's method has resulted in a well-balanced trees without chaining effects.

10.3   Evaluation with Respect to Requirements

In Section 3 the set of requirements for the intelligent media agent and for the specific application of a personal electronic program guide was given. An evaluation of the implementation comparing these requirements reveals which requirements are covered by the implementation presented. It is shown why specific requirements are not fulfilled, and how they can be met when extending the intelligent media agent.

10.3.1   General Remarks

In general, the test and evaluation of the prototype revealed the following results: By applying program filters, users get a personal TV which covers a reasonable amount of programs and which is restricted to the interesting offers. The introduction of priority parameters expressing the degree of interest allow for creating a hit list of favourite TV programs. Based on this structure, the electronic program guide application helps to find suitable programming in a faster and more convenient manner.

10.3.2   Requirements

Table 16 contains an overview showing which requirements have been fulfilled by the implementation. It can be seen that most requirements have been integrated into the design and implementation of the intelligent media agent.

Table 16: Summary of Requirements

Require-ment	Fulfilled (yes | partly | no)	Module Description
R1	yes	This requirement has been implemented within the multimedia run-time environment (MHEG, HTML/JAVA).
R2	yes	This requirement has been implemented within the multimedia run-time environment (MHEG, HTML/JAVA).
R3	yes	This requirement has been implemented as part of the Virtual Object Store.

Table 16: Summary of Requirements

Require-ment	Fulfilled (yes | partly | no)	Module Description
R4	partly	The architecture of the iMA provides for multiple agent applications. In case of the personal EPG there is one application which has been tested.
R5	yes	The agent run-time environment is capable of providing local and distributed processing by using the processing capabilities of MHEG and JAVA in client and server run-time environments.
R6	yes	Program descriptors have been introduced regarding multiple TV channels and program information.
R7	no	No mechanisms to dynamically update the personal EPG have been considered in this book. When discussing exceptional cases in the EPG application, it has been assumed that dynamic changes are not necessary to implement the core functionality of the iMA.
R8	yes	Program descriptors have been introduced containing pre-defined metadata about the TV channels and programs.
R9	yes	A pre-defined set of statistical information is used in the program descriptor.

Table 16: Summary of Requirements

Require-ment	Fulfilled (yes | partly | no)	Module Description
R10	yes	A genre tree has been defined as a base for a genre descriptor. By using this tree, a genre description is provided with fine granularity and can support multiple genres.
R11	yes	The quality of programming desired is given in the program descriptors.
R12	yes	Fuzzy sets have been introduced in order to describe fuzzy program attributes.
R13	partly	The user model has been defined by analysing user models taken from the television broadcasting area. Additionally, fuzzy information can be specified and attributes prioritized. It is not clear whether this model represents all user requirements.
R14	partly	Though the input of time of day, day of week and holidays is described, only the parameter "time of day" is used in the implementation of the EPG application. However, in case of using days and holidays, the algorithms would be similar.
R15	yes	User profiles contain attributes which are related to the corresponding attributes in the program descriptors.
R16	yes	Various user profile initialization modes have been defined and implemented.

Table 16: Summary of Requirements

Require-ment	Fulfilled (yes | partly | no)	Module Description
R17	yes	There are two versions of the EPG application implemented based on the portable MHEG and HTML/JAVA formats.
R18	yes	Identification and calculation methods provide a personal EPG (=multiple views to the TV program data)
R19	partly	There is no implementation dedicated to user groups and their administration. A user account can be shared by a group of users.
R20	yes	The administration interface has been implemented fulfilling the related requirement.
R21	yes	The required update modes have been introduced and implemented.
R22	yes	The personal EPG application augments traditional channel access, thereby guaranteeing that new interests can be explored in addition to the personal TV program.
R23	yes	Matching and clustering applications have been applied to the significant attributes of program descriptors and user profiles.
R24	yes	The results of matching and clustering are stable.
R25	yes	Personal preferences are specified in the user profiles which serve as input parameter to the matching and filtering operation.

Table 16: Summary of Requirements

Require-ment	Fulfilled (yes | partly | no)	Module Description
R26	yes	Certain exemplary exceptions have been defined. It has been shown how a rule-based system is used to detect and eliminate exceptions.
R27	yes	A user profile is a data structure which can be stored and transmitted to other set-top boxes.
R28	partly	It is possible to manipulate the decisions of the iMA by manipulating the TV program descriptors which are input to the matching and clustering process. It is recommended that the program description be done by independent authorities.
R29	yes	The used calculation scheme is a fixed function.
R30	partly	Privacy and security issues have been discussed. There is no implementation of a secure run-time environment for the intelligent media agent prototype.

10.4   Summary

Thusfar this requirements which have been fulfilled by the iMA system and personal EPG application have been examined. There are other requirements which have yet to be met: The need for multi-agent support (see R4) is proposed to be addressed by future extensions. Dynamic changes in the TV program schedule (see R7) have not been implemented because the basic functionality of the iMA and personal EPG does not require dynamic updates of the TV program schedule.

The lack of privacy and security (R29 and R30) is a major problem of the implementation. Any extensions should contain an implementation and test of security functions applied to user profile access and communication channels protection.

While implementing and testing the personal electronic program guide application, problems and potential improvements were encountered. For example, the installation and update of user profiles is too complex for people who are unfamiliar with complex user interfaces. Therefore, preferring major improvement, semi-automatic and automatic installation modes have been developed in Section 8.

The problem of small cluster generation arose while testing the process of clustering program and user profiles. This problem occurred because the "next neighbour" relation is transitive in most cases due to a large number of attributes (= dimensions) in the clustering process. To avoid the effect of small clusters, a multi-phase-clustering function has been developed. It allows for the definition of specific attributes as parameters for the first clustering loop (e.g. genre of programs). In subsequent loops, traditional clustering algorithms are applied.

Furthermore, a trade-off between the power of small footprint set-top boxes and the time-consumption of the matching and clustering algorithms was observed. Therefore, the following option for the generation of the personal electronic program guide is recommended: The generation of the guide should take place in an off-line mode in application server sites. However, privacy problems the arise.

In the matching and clustering process it has been found out that there are attributes in the user profile which seem to dominate. For example, the genre of a television program is an important filtering criteria. In order to reflect that importance, and as an improvement to the matching and clustering process, priority attributes should be introduced and attached to the attributes of user profiles, thereby making it possible to implement a usage model which regards the dominance of specific attributes in the user profile.

The matching and clustering operations are time-consuming. The introduction of attribute thresholds would pre-select important attributes and shorten the calculation of the personal EPG.

11 Conclusions

By presenting the concept and implementation of a personal electronic program guide as example application of an intelligent media agent, this work has examined how personal applications can be provided.

The primary focus of the first sections was the definition and classification of applications and intelligent agents. Based on a list of requirements the architecture of the intelligent media agent has been developed. This architecture incorporated the aspect of embedding the intelligent media agent standard interactive television such as DAVIC, and in multimedia programming environments such as MHEG together with their basic external and internal service interfaces. After that, the structure and specification of television program descriptors and user profiles has been discussed. Specific algorithms for comparing program descriptors and user profiles have been developed and tested. Clustering algorithms have been defined.

It has been recognized that the initialization and update of user profiles is complex. Therefore, various modes of initializing and updating user profiles have been introduced and discussed. By using statistical data and a user profile initialization table, an easy-to-use initialization mode have been provided. In the most advanced mode, the intelligent media agent observes all programs seen by an individual user and autonomously derives the user profile. It has been shown that clustering alone is not sufficient to automatically derive user profiles. Exceptions were found where the application of the clustering method failed (e.g., zapping, sleeping, visits). Based on this discussion, these exceptions were classified. A decision support system has been proposed in order to recognize and handle these exceptions.

A prototype including individual user initialization and static user profile adaptation has been implemented and tested. This implementation has shown that the intelligent media agent approach is suited to provide an application which regards the individual preferences of users. Parts of this book have been submitted to and considered by international standardization bodies (e.g., [Wittig 95a]). They have influenced the general definition of intelligent agent applications in the context of interactive television systems and the specific definition of open system interfaces which are necessary to implement personal electronic program guides.

11.1   Classification

The current intelligent Media Agent implementation is based on a static agent approach. There are no mobile script or agents. As part of the agent architecture, the intelligent Media Agent provides a generic user representation which can also interactively be used by other softbots (see Chapter 5). The current implementation takes into consideration the individual preferences of agent users, and uses reasoning mechanisms for exception handling. By applying adaptation rules the intelligent Media Agent can adapt its parameters, and partially learn from user feedback. The intelligent Media Agent is characterized as depicted in the spider diagram in Figure 49 (cf. Figure 5 of the preceding sections).

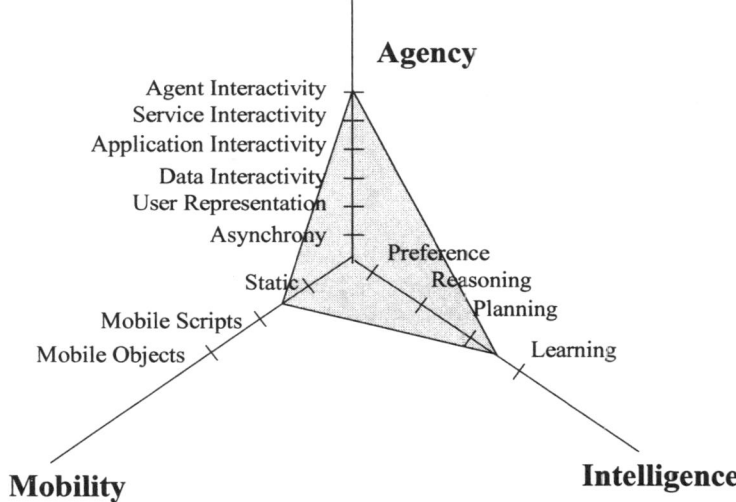

Figure 49: Classification of the iMA Implemented

11.2   Suggested Extensions

As part of the discussion of the results, future extensions can be seen. They address specific problems which were not addressed in Section 3 and remain unresolved.

11.2.1 Exploration of New Interests

An automatic update was defined and implemented to reflect changes in user interests. As a result, these changes are automatically detected and incorporated into the personal user profile. A continuous polling for new interests is an important design goal, the intelligent media agent being able to provide the following functionality:

In addition to the definition of the program hit list, a set of exploration programs can automatically offered. For example, the key of a remote control can be subdivided into keys for favourite programs and for programs offered for exploration (see Figure 50). Program types which have never been seen by the individual user before, or programs which are not among the group of favourite programs, are recognized and offered for exploration purposes to the user.

Figure 50: Exploration Channels

11.2.2 Content Based Retrieval

The attributes of channels, programs, scenes and cuts were regarded in the design of the intelligent media agent. The question of how attributes in the program descriptors can automatically be initialized is partly answered by referring to the research results in the area of content-based retrieval in VoD systems (see [Chen 94], [Dimitrova 94], and [Zhang 94]). Here, automatic computer-based classification and semantic description of scenes and clips is introduced. It is expected that similar methods can be applied to channels, scenes, and .

11.2.3 Information Broker

The intelligent media agent is based on a distributed approach. In order to provide for the sharing of knowledge in a multi-agent environment (i.e., among the intelligent media agents or between the intelligent media agents and other agents) a broker-based approach is proposed. The broker is a central module which provides an inter-agent communication framework for information and knowledge exchange among intelligent media agents and with other agents. Similar concepts have already been introduced in [Dilger 93]. For example, statistics agents are implemented based on the information broker. The statistic agent is responsible for collecting an anonymous television usage statistics by requesting accumulated usage lists from every intelligent media agent. In addition to the blackboard approach, the broker is an agent which can act autonomously based on own his intentions, decisions, autonomous acting. However, the broker is not authorized to instruct or manage intelligent media agents. Additionally, privacy issues must be taken into account in order to guarantee the anonymous collection of usage statistics.

11.2.4   Automated TV Program Director

An automated TV program director is an application which is responsible for automatic TV channel switching operations. For example, if a user prefers to exclude advertisement from his TV programs, the automated TV program director automatically switches the program if advertisement is transmitted. As soon as the program resumes or programming of interest to the user is again broadcast, current program may be interrupted. Users of the automated TV program director watch a virtual personalized channel instead of choosing among the other standard offerings.

11.2.5   Other Information Systems

The intelligent media agent can also be applied to other application domains.

Electronic Banking

The vision of using intelligent agents to support banking customers with personalized information systems has been described in [Wittig 96]. Instead of presenting standard banking information, user preferences should be regarded when the information is presented. Similar to the intelligent media agent presented in this book, the basic modules and functions could be applied to implement an intelligent banking application. The program descriptors and user profiles would need to be adapted to the corresponding application domain. An electronic marketplace has already been implemented using mobile intelligent agents (see [MMS 97]).

Personal Digital Newsletter

A personal digital newsletter (e.g., [Griwodz 97]) only contains that news which is of personal interest to the user. A possible genre tree for a personal digital newsletter is given in Figure 51. In addition to existing personal newsletter approaches, the intelligent media agent introduces the novel approach of using evaluative criteria to identify new programs of potential interest to the user. An example application has been implemented based on an object-oriented intelligent agent approach (see [Fiedler 97]).

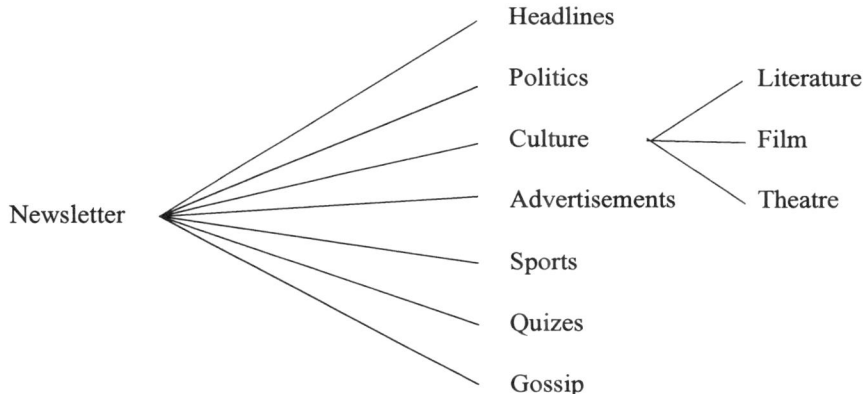

Figure 51: Genre Tree of a Personal Digital Newsletter

As a new functionality, news watching and filtering tools based on the intelligent agent paradigm would allow for automatically assisting users instead of manually describing favourite news content categories as in the Internet applications today. Thereby, the user interfaces of personalized news applications can be optimized.

Literature

[Agents 96]	Agents Inc.: *Firefly*. At www.agents-inc.com, 1996.
[Albayrak 93]	S. Albayrak, S. Bussmann: *Kommunikation und Verhandlungen in Mehragenten-Systemen*. In J. Müller (Ed.): *Verteilte Künstliche Intelligenz*. pp. 347-356, Springer, Heidelberg, 1993.
[Albayrak 92]	S. Albayrak: *Kooperative Lösung der Aufgabe Auftragsdurchsetzung in der Fertigung durch ein Mehr-Agenten-System auf der Basis des Blackboard-Modelles*. PhD. Thesis, Technical University of Berlin, 1992.
[Albers 63]	J. Albers: *The Interaction of Color*. Yale University Press, New Haven, CT, 1963.
[Alberico 90]	R. Alberico, M. Micco: *Expert Systems for Reference and Information Retrieval*. Meckler. London, 1990.
[Arbib 87]	M.A. Arbib and A.R. Hanson, (Eds.): *Vision, Brain, and Cooperative Computation*. MIT Press/Bradford Books, MA, 1987.
[Backhaus 87]	K. Backhaus, B. Erichson, W. Plinke, C. Schuchard-Ficher, R. Weiber: *Multivariate Analysemethoden*. pp. 115-159, Springer, Heidelberg, 1987.
[Berkom 94]	DEC, GMD Fokus, Grundig Multimedia Solutions, IBM ENC, Technical University of Berlin PRZ: *BERKOM Globally Accessible Services: System Specification 1.0*. BERKOM, Berlin, May 1994.
[Benyon 87]	D. Benyon, D. Murray: *Experience with adaptive interfaces*. Computer Journal 31(5), pp. 465-473, October 1987.
[Bertin 94]	C. Bertin: *Eurescom IMS1 Projects (Integrated Multimedia Services at about 1 MBit/s*. 2nd International Workshop on Advanced Teleservices and High-Speed Communication Architectures (IWACA), Heidelberg, 1994.
[Bezdek 93]	J.C. Bezdek: *Fuzzy Models - What Are They and Why?* IEEE Transactions on Fuzzy Systems, No. 1, 1993.
[Bingham 93]	J. Bingham, W.Y. Chen: *NEXT from T1 to ADSL and Vice Versa*. Amati Communication and Bellcore Contribution, ANSI T1E1.4/93-178, August 1993.
[Bouron 92]	T. Bouron, A. Collinot: *SAM: A Model to Design Complex Computational Social Agents*. ECAI-92, pp. 239-243, 1992.
[Boy 91]	G.A. Boy: *Intelligent Assistant Systems*. Academic Press, San Diego, 1991.
[Brenner 97]	W. Brenner, R. Zarnekow, H. Wittig: *Intelligente Softwarea-*

genten. Springer, Heidelberg, 1997.

[Breunig 96]   C. Breunig: *Pay Radio - ein neues Hörfunkangebot*. Media Perspektiven 7/96, pp.375-385, Frankfurt, 1996.

[Brooks 86]   R.A. Brooks. *A Robust Layered Control System for a Mobile Robot*. IEEE Journal of Robotics and Automation, 2 (1), pp.14-23, 1986.

[Brooks 90]   R.A. Brooks. *The Behavior Language; User's-Guide*. Technical report. Memo 1227, MIT, Boston, 1990.

[BT 97]   British Telecommunications: *Movie Recommendation System (MORSE)*. At www.labs.bt.com/morse/morse0-q, August 1997.

[BTUC 97]   Brandenburgische Technische Universität Cottbus: *WWW-Watchcat*. Agent example applications developed within the lecture "Multimedia Systems" given by H.Wittig, Brandenburg Technical University, Cottbus, 1997.

[Canary 93]   D.J. Canary, B.H. Spitzberg: *Loneliness and media gratification*. Communications Research, Vol. 20, No. 6, pp. 800-821, 1993.

[Carrol 91]   J.M. Carrol (Ed.): *Design Interaction: Psychology at the Human-Computer Interface*. Cambridge University Press, NY, 1991.

[Chen 94]   S.S. Chen, S.S. Shen: *Multimedia Computing for Digital Libraries*. Newsletter of the Technical Committee on Multimedia Computing of IEEE Multimedia, June 1994.

[Chen 94]   W.Y. Chen, D.L. Waring: *Applicability of ADSL to Support Video Dial Tone in the Copper Loop*. IEEE Communications, May 1994.

[Chiari 95]   L. Chiariglione: *The Digital Audio-Visual Council*. IEEE Computer Society Multimedia Newsletter, H. Wittig (Editor in Chief), December 1995.

[Chin 91]   D. Chin: *Intelligent interfaces agents as agents*. Intelligent User Interfaces, ACM Press, 1991.

[Chow 92]   J.S. Chow, J.C. Tu, J.M. Cioffi: *A discrete multitone transceiver system*. IEEE JSAC, Vol.9, No. 6, pp. 895-908, August 1992.

[Chua 94]   T.-S. Chua, S.-K. Lim, H.-K. Pung: *Content-based retrieval of Segmented Images*. Proceedings of the 2nd ACM Multimedia Conference, San Francisco, CA, October 1994.

[Clancey 87]   W.J. Clancey: *Knowledge Based Tutoring: The GUIDON program*. MIT Press, Cambridge, MA, 1987.

[Coen 94]   M.H. Coen: *Sodabot - A Software Agent Environment and Construction System*. Extended Abstract, MIT AI Lab, September

1994.

[Corkill 91]   D. Corkill: *Blackboard Systems*. AI Expert, Vol. 9, No. 91, pp. 41-47, 1991.

[Cossmann 96]   H. Cossmann, C. Griwodz, G. Grassel, M. Pühlhöfer, M. Schreiber, R. Steinmetz, H. Wittig, L. Wolf: *GLASS — A Distributed MHEG-Based Multimedia System*. Multimedia Computing and Networking (MNCN), San Jose, January 1996.

[Cossmann 95]   H. Cossmann, C. Griwodz, G. Grassel, M. Pühlhöfer, M. Schreiber, R. Steinmetz, H. Wittig, L. Wolf: *Interoperable iTV Systems based on MHEG*. International COST237 Workshop on Multimedia Transport and Teleservices, Copenhagen, December 1995.

[Cruse 93]   H. Cruse, U. Muller-Wilm, J. Dean: *Artificial Neural Nets for controlling a 6-legged walking system*. In J.-A. Meyer, H.L. Roitblatt, and S.W. Wilson, (Eds.): *From Animals to Animats2*. Proceedings of the 2nd International Conference on Simulation of Adaptive Behavior, pp. 52-60, 1993.

[CDI 92]   Phillips Inc.: *The CD-I Handbook Series*. Addison-Wesley, 1992.

[CMU 97]   Carnegie Mellon University: *FAQ: Fuzzy Logic and Fuzzy Expert Systems*. At www.cs.cmu.edu/Groups/AI/html/faqs/fuzzy/part1/faq-doc-2.html, August 1997.

[CMU 96]   Carnegie Mellon University: *Movie Recommendation Engine*. At phoebe.dws.acs.cmu.edu/cgi-bin/movie, 1996.

[Dahm 96]   I. Dahm, C. Griwodz, M. Schreiber, J. Winckler: *Going Multimedia in Interactive TV systems - pitfalls of MHEG and JAVA compared*. iTV Conference, Edinborough, 1996.

[Darschin 95]   W: Darschin, B. Frank: *Tendenzen im Zuschauerverhalten 1994*. Media Perspektiven 4/95, pp. 154 - 171, Frankfurt, 1995.

[Darschin 96]   W. Darschin, B. Frank: *Tendenzen im Zuschauerverhalten 1995*. Media Perspektiven 4/96, pp. 174-185, Frankfurt, 1996.

[Dilger 93]   W. Dilger, S. Kassel: *Sich selbst organisierende Fertigungsprozesse als Möglichkeit zur flexiblen Fertigungssteuerung*. In J. Müller (Ed.): *Verteilte Künstliche Intelligenz*. pp. 347-356, Springer, Heidelberg, 1993.

[Dimitrova 94]   N. Dimitrova, F. Golshani: *Rx for Semantic Video Database Retrieval*. Proceedings of the 2nd ACM Multimedia Conference, San Francisco, October 1994.

[DAVIC 95a]   Digital Audio-Visual Council (DAVIC): *Part 1: Description of DAVIC Functionalities*. DAVIC Specification Version 3.3, Co-authored by H.Wittig, Cagliari, March 1995.

[DAVIC 95b]   Digital Audio-Visual Council (DAVIC): *Part 5: Set-Top Unit Architecture and Application Programming Interface.* DAVIC Specification 1.0 Revision 2.0, Co-authored by H.Wittig, Cagliari, May 1995.

[DAVIC 95c]   Digital Audio-Visual Council (DAVIC): *Part 9: Information Representation.* DAVIC 1.0 Specification 1.0 Revision 2.0, Co-authored by H.Wittig, Cagliari, May 1995.

[DAVIC 95d]   Digital Audio-Visual Council (DAVIC): *DAVIC 1.0 Revision 5.0,* Co-authored by H.Wittig, Cagliari, December 1995.

[DAVIC 96]   Digital Audio-Visual Council (DAVIC): *Meeting Minutes of the DAVIC Software Portability Ad-Hoc Group,* Chaired by H.Wittig, Heidelberg, January 1996.

[DF1 97]   DF1 GmbH & Co. KG: *DF1 - Das digitale Fernsehen.* At www.df1.de, August 1997.

[DVI 92]   Deutsches Video Institut e.V.: *Markt- und Medienanalyse.* Berlin, 1990, 1991, 1992.

[DVB 94]   European Broadcasting Union / European Telecommunication Standards Institute: *Digital broadcasting systems for television, sound and data services; Specification for Services Information (SI) in Digital Video Broadcasting (DVB) Systems.* ETS 300 486, October 1994.

[Erlandsen 87]   J. Erlandsen, J. Holm: *Intelligent Help Systems.* Information and Software Technology, 29(3), pp. 115-121, 1987.

[Ehrmantraut 95]   M. Ehrmantraut: *Verfahren zur benutzerspezifischen Informationsrepräsentation in interaktiven Fernsehsystemen.* Diploma Thesis, Supervised by T. Härder (University of Kaiserslautern) and H.Wittig (IBM ENC), University of Kaiserslautern, IBM ENC Heidelberg, 1995.

[Ehrmantraut 96]   M. Ehrmantraut, T. Härder, H. Wittig, R. Steinmetz: *The Personal Electronic Program Guide — Towards the Pre-selection of Individual TV Programs.* 5th International Conference on Information and Knowledge Management (CIKM), Toronto, 1996.

[El-Ham 89]   A. El-Hamdouchi, P. Willett: *Comparison of hierarchical agglomerative clustering methods for document retrieval.* Computer Journal, 32(3), pp. 220-227, 1989.

[Feier 96]   S. Feierabend, T. Windgasse: *Was Kinder sehen.* Media Perspektiven 4/96, pp. 196-194, Frankfurt, 1996.

[Fiedler 96]   J. Fiedler, T. Bluhm: *Metasearch Engine.* Prototype implementation as result of Internship supervised by H.Wittig, Heidelberg, 1996.

[Fiedler 97]   J. Fiedler: *Mobile Agenten im Umfeld personalisierter Nach-richtendienste.* Diploma Thesis. Supervised by H. König, I.Beier, H.Wittig, R.Zarnekow, Brandenburg Technical University, Cottbus, 1997.

[Finin 92]   T. Finin, J. Weber, G. Wiederhold, M. Genesereth, R. Fritzson, D. McKay, J. McGuire, P. Pelavin, S. Shapiro, C. Beck: *Specification of the KQML Agent-Communication Language.* Enterprise Integration Technologies, Palo Alto, CA, Technical Report EIT TR 92-04, 1992.

[Fisk 96]   D. Fisk: *An application of social filtering to movie recommendation.* British Telecom Technology Journal. Vol. 14, No. 4, October 1996.

[Gabbe 94]   J.D. Gabbe, A. Ginsberg, B.S. Robinson: *Towards Intelligent Recognition of Multimedia Episodes in Real-Time Applications.* Proceedings of the 2nd ACM Multimedia Conference, San Francisco, October 1994.

[Genesereth 94] M.R. Genesereth, S.P. Ketchpel: *Software Agents.* Communications of the ACM, Vol. 37, No. 7, July 1994.

[Genesereth 92] M.R. Genesereth, R.E. Fikes: *Knowledge Interchange Format Version 3 Reference Manual.* Logic-92-1, Stanford University Logic Group, 1992.

[Gerhards 96]   M. Gerhards, A. Grajczyk, W. Klingler: *Programmangebote und Spartennutzung im Fernsehen 1995.* Media Perspektiven 11/96, pp. 572-577, Frankfurt, 1996.

[Gleich 95]   U. Gleich: *Zuschauermotivation für den Fernsehkonsum.* Media Perspektiven 4/95, pp. 186-191, Frankfurt, 1995.

[Göllner 95]   V. Göllner: *Analyse charakteristischer Benutzermerkmale zur Erstellung und Integration von Standard-Regeln für intelligente Medienagenten.* Diploma Thesis. Supervised by W. Brill (Berufsakademie Mannheim) and H. Wittig (IBM ENC), Berufsakademie Mannheim, 1995.

[Greif 94]   I. Greif: *Desktop Agents in Group-Enabled Products.* Communications of the ACM, Vol. 37, No. 7, July 1994.

[Grajczyk 96]   A. Grajczyk, O. Zöllner: *Fernsehverhalten und Programmpräferenzen älterer Menschen.* Media Perspektiven 11/96, pp. 577-588, Frankfurt, 1996.

[Griwodz 97]   C. Griwodz, M. Liepert: *Personalised News on Demand: The HyNoDe Server.* In Proc. of the 4th International Workshop on Interactive Distributed Multimedia Systems and Telecommunication Services (IDMS'97), pg. 241-250, 1997.

[Henrion 88]   G. Henrion, A. Henrion, R. Henrion: *Beispiele zur Datenana-*

	lyse mit BASIC-Programmen. Deutscher Verlag der Wissenschaften, Berlin, 1988.
[Hutchinson 94]	J.M. Hutchinson: *A Radial Basis Function Approach to Financial Time Series Analysis.* PhD. Thesis, MIT AI Laboratory, Boston, 1994.
[IBM 96]	*IBM Intelligent Agent Strategy White Paper.* http://activist.gpr.ibm.com:81/whitepaper/proz.htm, 1996
[ISO 10918]	ISO/IEC International Standard 10918:1992: *Information Technology – Digital Compression and Coding of Continuous-Tone Still Images (JPEG).* 1992.
[ISO 11172]	ISO/IEC International Standard 11172:1992: *Information Technology – Coding of Moving Pictures and Associated Audio for Digital Storage Media up to about 1.5 Mbit/s (MPEG).* 1992.
[ISO 11544]	ISO/IEC International Standard 11544:1992: *Information Technology – Digital Compression and Coding of Bi-level Images (JBIG).* 1992.
[ISO 135221]	ISO/IEC Working Draft 13522-1:1994: *Information Technology – Coded Representation of Multimedia and Hypermedia Information Objects (MHEG) – Part 1: Base Notation (ASN.1).* June 1993.
[ISO 135225]	ISO/IEC Committee Draft 13522-5: *Information Technology - Coding of Multimedia and Hypermedia Information - Part 5.* Committee Draft, Multimedia and Hypermedia Experts Group (MHEG), Co-authored by H.Wittig, Tokyo, July 1995.
[ISO 135521]	ISO/IEC Committee Draft 13552-1:1993: *Information Technology – MPEG-2 Digital Storage Media Command and Control Extension (DSM-CC).* November 1994.
[ITUT 170]	ITU-T Draft Recommendation T.170: *Audiovisual Interactive (AVI) Systems – General Introduction, Principles, Concepts and Models.* Second Revision, Geneva, CH, November 1993.
[Jäckel 93]	M. Jäckel: *Fernsehwanderungen. Eine empirische Untersuchung zum Zapping.* R. Fischer, Munich, 1993.
[Jarvis 73]	R. A. Jarvis, E. A. Patrick: *Clustering Using a Similarity Measure Based on Shared Near Neighbors.* IEEE Transactions on Computers, Vol. C-22, No. 11, November 1973.
[Kautz 94]	H. Kautz, B. Selman, M. Coen, S. Ketchpel: *An experiment in the design of software agents.* Proceedings of the 12th National Conference on Artificial Intelligence, AAAI-93, Washington D.C., 1994.
[Kaelbling 87]	Leslie Pack Kaelbling: *An architecture for intelligent reactive systems.* In Michael P. Georgeff and Amy L. Lansky, Eds., *Rea-*

soning about Actions and Plans, pp. 395-410, Morgan Kaufmann, 1987.

[Kehoe 85]   C. Kehoe: *Interfaces and expert systems for on-line retrieval.* Online Review, 9(6), pp. 489-505, 1985.

[Kosko 92]   B. Kosko. *Neural Networks and Fuzzy Systems: A Dynamical Systems Approach to Machine Intelligence.* Prentice Hall, Englewood Cliffs,NJ, 1989.

[Kleinegees 94]   U. Kleinegees, C. Krueger-Hemmer, G. Kyi, C. Weller: *Im Blickpunkt: Kultur in Deutschland: Zahlen und Fakten.* Statistisches Bundesamt, Wiesbaden, 1994.

[Krüger 96a]   U.M. Krüger: *Tendenzen in den Programmen der großen Fernsehsender 19985-1995.* Media Perspektiven 8/96, pp. 418-440, Frankfurt, 1996.

[Krüger 96b]   U.M. Krüger: *Gewalt in von Kindern genutzten Fernsehsendungen.* Media Perspektiven 3/96, p. 121, Frankfurt, 1996.

[Laurel 91]   B. Laurel: *The Art Of Human-Computer Interface Design.* Addison-Wesley, Reading NY, 1991.

[Lin 93]   C.A. Lin: *Modelling the gratification-seeking process of television viewing.* Human Communication Research, Vol. 20, No. 2, pp. 224-244, 1993.

[Lloyd 90]   P.R. Lloyd: *Creating Knowledge-based multimedia applications.* Colloquium of the Committees C4/C5/E14 of the IEE Computing and Control Division, London, 1990.

[Lochte 93]   Robert H. Lochte: *Interactive Television and Instruction: a guide to technology, technique, facilities, design and classroom management.* Educational Technology Publications, Englewood Cliffs,NJ, 1993.

[Maes 94]   P. Maes: *Agents that Reduce Work and Information Overload.* Communications of the ACM, Vol. 37, No. 7, July 1994.

[Maes 93a]   P. Maes, R. Kozierok: *Learning Interface Agents.* Proceedings of the 11th National Conference on Artificial Intelligence (AAAI), July 1993.

[Maes 93b]   P. Maes: *Evolving Agents for Personalized Information Filtering.* Proceedings of the 9th Conference on AI for Applications, CAIA'93, 1993.

[Markus 93]   A. Markus: *Future Directions in Advanced Interface Design.* In Communication with Virtual Worlds, Springer, Heidelberg, 1993.

[Maturana 87]   H.R. Maturana, F.J. Varela: *The Tree of Knowledge: The Biological roots of Human Understanding.* Shamhala Press, Boston, 1987.

[McFarland 92]	D. McFarland: *Autonomy and self-sufficiency in robots.* Technical Report, VUB AI Lab Memo 92-3, 1992.
[Media 94a]	Media Control: *Top-20 Datenmaterial.* Media Control, Baden-Baden, Germany, 1994.
[Media 94b]	Media Perspektiven: *Daten zur Mediensituation in Deutschland 1994.* Frankfurt, 1994.
[Medior 95]	Medior Inc.: *Movie Select: The Intelligent Guide over 44,000 Videos.* CD-ROM, Paramount Interactive, 1995.
[Menzer 95]	F. Menzer, R.K. Belew, W. Willuhn: *Artificial Life Applied to Adaptive Information Agents.* Technical Report, Communications Technology Lab, ETH Zürich, 1995.
[Meyer 95]	T. Meyer-Boudnik, W. Effelsberg: *MHEG - An Interchange Format for Interactive Multimedia Presentations.* IEEE Multimedia Magazine, Summer 1995.
[Michel 97]	S. Michel: *Entwicklung einer personalisierbaren Online-Zeitschrift.* Practice Work, supervised by K. Meißner, T. Jörding, H. Wittig, Technical University of Dresden, 1997.
[Mucha 92]	H.J. Mucha: *Clusteranalyse mit Mikrocomputern.* Akademie-Verlag, Berlin, 1992.
[MMS 96a]	Multimedia Software GmbH Dresden: *iTV -Feldversuch der Deutschen Telekom AG in Nürnberg.* Flyer, 1996.
[MMS 96b]	Multimedia Software GmbH Dresden: *Quelle Tele-Shopping Application.* Flyer, 1996.
[MMS 97]	Multimedia Software GmbH Dresden: *PET- Personal Electronic Trader.* Flyer, CeBit, Hannover, 1997.
[Negroponte 94]	N. Negroponte: *Keynote Address.* IEEE 1st International Conference on Multimedia Computing and Systems, Boston, May 1994.
[Newell 62]	A. Newell: *Some problems in basic organizations in problem solving.* In M.C. Yovits, G.T. Jacobi, G.D. Goldstein (Eds.), Conference of Self-Organizing Systems, Washington D.C., pp. 189-198, 1962.
[Oehm 96]	E. Oehmichen, E. Simon: *Fernsehnutzung, politisches Interesse und Wahlverhalten.* Media Perspektiven 11/96, pp. 562-571, Frankfurt, 1996.
[OMalley 86]	C.E. O'Malley: *Helping users help themselves.* User Centered System Design: New Perspectives on Human-Computer Interaction. Erlbaum Pub., 1986.
[OMG 91.12.1]	Object Management Group: *Common Object Request Broker: Architecture and Specification.* OMG 91.12.1, 1991.

[Opasch 94]   Opaschewski: *Fernsehkonsum: Fakten und Trends.* BAT
                  Freizeitinstitut, 1994.

[Orwant 96]   J. Orwant: *For want of a bit the user was lost: Cheap user mod-*
                  elling. IBM Systems Journal. Vol. 35, Nos. 3&4, 1996.

[Philips 88]   Philips International Inc.: *Compact Disc-Interactive.* McGraw-
                  Hill, New York, 1988.

[Pfeiffer 92]   R. Pfeiffer, P. Verschure: *Distributed Adaptive Control: A Par-*
                  adigm for Designing Autonomous Agents. In F.J. Varela and P.
                  Bourgine, Eds., *Toward a Practice of Autonomous Systems.*
                  Proceedings of the First European Conference of Artifical Life,
                  pp. 21-30, MIT Press/Bradford Books, 1992.

[Potts 94]   R. Potts, D. Sanchez: *Television viewing and depression: no*
                  news is good news. Journal of Broadcasting and Electronic
                  Media 38, No. 1, pp. 79-90, 1994.

[Price 93]   R. Price: *An introduction to future international standard for*
                  hypermedia object interchange. Proceedings of the 1st ACM
                  Multimedia, Anaheim CA, August 1993.

[Rasmussen 93]   E. Rasmussen: *Clustering algorithms.* W. Frakes and R. Beat-
                  Yates (Eds.), Information Retrieval - Data Structures and Algo-
                  rithms, Chapter 16, Prentice Hall, Englewood-Cliffs NJ, 1993.

[Reimers 95a]   U. Reimers: *The Digital Video Broadcasting Project.* IEEE
                  Computer Society Multimedia Newsletter, H. Wittig (Ed.),
                  December 1995.

[Reimers 95b]   U. Reimers: *Digital Video Broadcasting (DVB).* March 1995.

[Resnick 94]   P. Resnick, N. Iacovou, M. Suchak, P. Bergstrom, J. Riedl:
                  GroupLens: An Open Architecture for Collaborative Informa-
                  tion Filtering of Netnews. Proceedings of the ACM Conference
                  on Computer Supported Collaborative Work. Chapel Hill NC,
                  1994.

[Riecken 94]   D. Riecken: *An Architecture of Integrated Agents.* Communica-
                  tions of the ACM, Vol. 37, No. 7, July 1994.

[Riley 91]   G. Riley: *CLIPS: An Expert System Building Tool.* Proceedings
                  of the Technology 2001 Conference, Washington D.C., Novem-
                  ber 1991.

[Rosenberg 81]   S. Rosenberg: *Expert systems and the design of powerful user*
                  interfaces. Information and Community. An Alliance for
                  Progress, Proceedings of the 44th American Society for Infor-
                  mation Science Annual Meeting, pp. 285-297, 1981.

[Rouse 87]   W.B. Rouse, N.D. Geddes, R.E. Curry: *An Architecture for*
                  Intelligent Interfaces: Outline of an Approach to Supporting
                  Operators of Complex Systems. Human Computer Interaction.

Vol. 3, Erlbaum Pub., London, 1987.

[Sack 94]   W. Sack, M. Davis: *IDIC: Video Sequences from Story Plans and Content Annotations.* Proceedings of the 1st IEEE International Conference on Multimedia Computing and Systems, Boston, May 1994.

[Salton 87]   G. Salton, M.J. McGill: *Information Retrieval - Grundlagen für Informationswissenschaftler.* McGraw-Hill NY, 1987.

[Schmitz 93]   B. Schmitz, C. Alsdorf, F. Sang, K. Tasche: *Der Einfluß psychologischer und familiaerer Rezipientenmerkmale auf die Fernsehmotivation.* Rundfunk und Fernsehen, 41/1, pp. 5-19, 1993.

[Seel 89]   N. Seel: *Agent Theories and Architecture.* PhD. thesis, Surrey University, Gilford, UK, 1989.

[Shardlow 90]   N. Shardlow: *Action and agency in cognitive science.* Master's thesis. Dept. of Psychology, University of Manchester, UK, 1990.

[Sheth 94]   B. D. Sheth: *A Learning Approach to Personalized Information Filtering.* M.I.T. Media Laboratory, Technical Report 1, 1994.

[Siemens 95]   Siemens: *Private Communication.* CeBit, Hannover, 1995.

[Steels 94]   L. Steels: *The artifical life roots of artificial intelligence.* Artificial Life Journal, January 1994.

[Steinmetz 95] R. Steinmetz, K. Nahrstedt: *Multimedia Systems.* Prentice Hall, 1995.

[Steinmetz 96] R. Steinmetz, H. Wittig: *Multimedia.* Chapter of BIT Special Edition on Multimedia System, Tokyo, 1996.

[Stone 86]   M. Stonebraker, C. Hauson, E. Hong: *The design of the POSTGRES Rule System.* Proceeding of the Data Engineering Conference, Los Angeles, 1986.

[Sung 94]   J.-J. Sung, M.-Y. Huh, H.-J. Kim, J.-H. Hahm: *Hypermedia Information Retrieval Systems Using MHEG Coded Representation in a Networked Environment.* 2nd International Workshop on Advanced Teleservices and High-Speed Communication Architectures (IWACA), Heidelberg, 1994.

[UMN 96]   University of Minnesota at Minneapolis: *GroupLens.* At www.cs.umn.edu/Research/GroupLens, 1996.

[Vazirian 95]   Sh.Vazirian: *Entwicklung eines Regelwerks fuer intelligente Assistenten in interaktiven Fernsehsystemen am Beispiel BERKOM GLASS.* Diploma Thesis, supervised by U. Haisch (FHT Mannheim) and H. Wittig (IBM ENC), Fachhochschule Technik, Mannheim, 1995.

[Vorderer 94]   P. Vorderer: *Involvementverläufe bei der Rezeption von Fernse-
                  hfilmen.* L. Bosshart, W. Hoffmann-Riem (Eds.), Medienlust
                  und Mediennutz, München, Ölschläger, pp. 333-342, 1994.

[Vorhees 94]   E.M. Vorhees: *Software Agents for Information Retrieval.*
                  Working Notes of the AAAI Spring Symposium on Software
                  Agents, pp. 126-129, AAAI Press, Menlo Park CA, March
                  1994.

[Weiler 97]   S. Weiler: *Computernutzung und Fersehkonsum von Kindern.*
                  Media Perspektiven 1/97, pp. 43-53, Frankfurt, 1997.

[Wheeler 94]   M. Wheeler: *For Whom the Bell Tolls? The Roles of Represen-
                  tation and Computation in the Study of Situated Agents.* Techni-
                  cal Report, School of Cognitive and Computational Sciences,
                  University of Sussex, Brighton, UK, 1994.

[Willet 88]   P. Willett: *Recent trends in hierarchic document clustering: A
                  critical review.* Information Processing and Management, No.
                  24, pp. 577-597, May 1988.

[Wiles 94]   R. Wiles, S. Wavered: *Service Requirements for Video-on-
                  Demand ATM Transport.* ATM Forum Contribution 94-688,
                  July 1994.

[Wilson 95]   L. Wilson: *Intelligent Agents: A Primer.* IBM Personal Systems
                  Journal, September/October 1995.

[Wittig 95a]   H. Wittig: *Personal EPG's as DAVIC Applications.* 7th DAVIC
                  Meeting, Application Technical Committee, London, March
                  1995.

[Wittig 95b]   H. Wittig, C. Griwodz: *Intelligent Media Agents for Interactive
                  Television Systems.* IBM ENC Technical Report 43.9501, Hei-
                  delberg, January 1995.

[Wittig 95c]   H. Wittig, C. Griwodz: *Intelligent Media Agents in Interactive
                  Television Systems.* International Conference on Multimedia
                  Computing and Systems '95, Boston, May 1995.

[Wittig 95d]   H. Wittig, G. Grassel: *MHEG as Interoperable Application
                  Format.* Answer to the 2nd DAVIC Call for Proposals, Cagliari,
                  May 1995.

[Wittig 95e]   H. Wittig, C. Griwodz: *Intelligent Interactive Television Sys-
                  tems.* Dortmunder Fernsehseminar, Dortmund, September
                  1995.

[Wittig 96]   H. Wittig: *Electronic Banking.* In Schoop, Glowalla (Eds.), Per-
                  spektiven multimedialer Kommunikation. Springer, Heidel-
                  berg, 1996.

[Wolf 94]   L.C. Wolf, R.G. Herrtwich: *The System Architecture of the Hei-
                  delberg Transport System.* ACM Operating Systems Review,

Vol. 28, No. 2, pp. 51-64, April 1994.

[Wolf 95]   L. Wolf: *Resource Management for Distributed Multimedia Systems*. PhD. Thesis, Technical University of Chemnitz-Zwickau, 1995.

[Zadeh 65]   L. Zadeh: *Fuzzy Sets*. Information and Control, No. 8, 1965.

[Zhang 94]   H. Zhang, Y. Gong, S.W. Smoliar, S.Y. Tan: *Automatic Parsing of News Video*. Proceedings of the 1st IEEE International Conference on Multimedia Computing and Systems, Boston, May 1994.

[Zimmer 96]   J. Zimmer: Pay TV: *Durchbruch im digitalen Fernsehen?* Media Perspektiven 7/96, Frankfurt, 1996.

[Zhang 93]   L. Zhang, S. Deering, D. Estrin, S. Shenker, D. Zappala: *RSVP: A New Resource ReSerVation Protocol*. IEEE Network, September 1993.

Abbreviations

AAL	Asynchronous Transfer Mode Adaptation Layer
ADSL	Asymmetric Digital Subscriber Line
AGF	Arbeitsgemeinschaft Fernsehforschung
AMS	Audiovisual and Multimedia Services
ANSI	American National Standards Institute
ARD	Arbeitsgemeinschaft der Rundfunkanstalten Deutschlands
ATM	Asynchronous Transfer Mode
AVI	Audio Video Interleaved
BERKOM	Berliner Kommunikationssystem
BTX	Bildschirmtext
CBR	Constant Bit-Rate
CCIR	International Radio Consultative Committee
CCITT	International Telegraph and Telephone Consultative Committee
CD	Committee Draft / Compact Disc
CD-ROM	Compact Disk - Read Only Memory
CNN	Cable News Network
CORBA	Common Object Request Broker Architecture
CSCW	Computer Supported Cooperative Work
CV	Cybervision
DAB	Digital Audio Broadcasting
DAT	Discrete Attribute Type
DAVIC	Digital Audio-Video Council
DIN	Deutsche Institut für Normung
DIS	Draft International Standard
DSM-CC	Digital Storage Media Command and Control
DSOM	Distributed System Object Model
DVB	Digital Video Broadcasting
EBU	European Broadcasting Union
ENC	European Networking Center
EPG	Electronic Program Guide
ETSI	European Telecommunication Standards Institute
FAQ	Frequently Asked Questions
FTP	File Transfer Protocol
FTTC	Fibre To The Curb

FTTH	Fibre To The Home
GfK	Gesellschaft für Konsumforschung
GLASS	Globally Accessible Services
HAT	Hierarchy Attribute Type
HDTV	High Definition Television
HyTime	Hypertext Multimedia Extension
HTML	Hypertext Mark-up Language
IBM	International Business Machines
IEC	International Electrotechnical Commission
IEEE	Institute of Electrical and Electronics Engineers
IETF	Internet Engineering Task Force
iMA	Intelligent Media Agent
IP	Internet Protocol
IS	Industry Standard
ISDN	Integrated Services Digital Network
ISO	International Standardization Organization
ITU	International Telecommunications Union
iTV	Interactive Television
IV	Interactive Vision
JBIG	Joint Bi-Level Images Experts Group
JTC	Joint Technical Committee
JPEG	Joint Photographic Experts Group
KIF	Knowledge Interchange Format
KQML	Knowledge Query and Manipulation Language
LAN	Local Area Network
LNB	Low Noise Block
MAN	Metropolitan Area Network
MAT	Multi-Attribute Type
MIME	Multipurpose Internet Mail Extensions
MHEG	Multimedia and Hypermedia Experts Group
MoD	Movie on Demand
MPEG	Moving Pictures Experts Group
MTU	Maximum Transfer Unit
MTV	Music Television
NAT	Number Attribute Type
NCSA	National Center for Supercomputing Applications
NoD	News on Demand

NVoD	Near-Video-on-Demand
OSI	Open System Interconnection
PDU	Protocol Data Unit
PIN	Personal Identification Number
PS	Program Stream
PVC	Permanent Virtual Circuits
QoS	Quality of Service
ROM	Read Only Memory
RFC	Request For Comments
RPC	Remote Procedure Call
SC	Subcommittee
SDU	Service Data Unit
SNMP	Simple Network Management Protocol
SoD	Services on Demand
SOM	System Object Model
SVC	Switched Virtual Circuits
STB	Set-Top Box
STT	Set Top Terminal
STU	Set Top Unit
TCP	Transmission Control Protocol
TV	Television
UNI	Universal Network Interface
VBR	Variable Bit-Rate
VC	Virtual Circuit
VCR	Video Recorder
VoD	Video-on-Demand
VR	Virtual Reality
WAN	Wide Area Network
WD	Working Draft
WG	Working Group
WWW	Worldwide Web
ZDF	Zweites Deutsches Fernsehen

Appendix A: Genre Trees

In the following, the genre tree as part of the user and movie profile database are given. The root genres are shown in Figure 52.

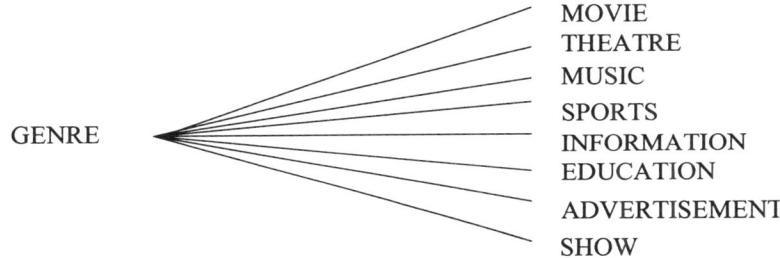

GENRE

MOVIE
THEATRE
MUSIC
SPORTS
INFORMATION
EDUCATION
ADVERTISEMENT
SHOW

Figure 52: Root Genres

Figure 53 indicates that subgenre of "Theatre" and "Music" overlap. The necessity of overlaps is shown in the following example. There are two programs: Program A is a movie biography about the composer Johann Strauß. This movie necessarily needs to include excerpts from his operettas. Program B contains the operetta "Batman". composed by Johann Strauß. The classification of both programs is as follows:

Table 17: Programs Examples[a]

Genre	Program A	Program B
MOVIE->Theatre	70	0
MOVIE->Music	65	0
THEATRE->Operetta	0	100
MUSIC->Operetta	100	100

a. The attribute values range from 0 to 100.

The classification in the genre tree reflects the movie-type of program A. Program B is classified as an operetta performance. The music of both program A and B is operetta music.

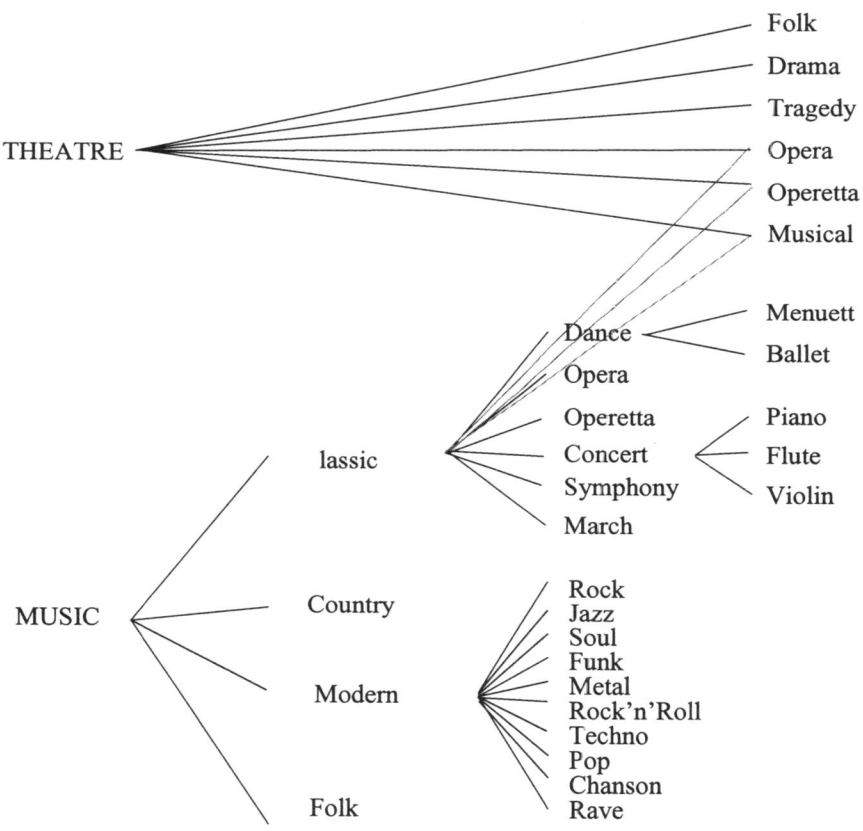

Figure 53: Subgenres of Genre "Theatre" and "Music"

The definition of the genres "Sports", "Information", "Education" and "Shows" is given in Figure 54-57. If required, the genre tree can be extended by inserting new subgenres or references to existing parts of the genre tree.

Figure 54: Subgenres of Genre "Sports"

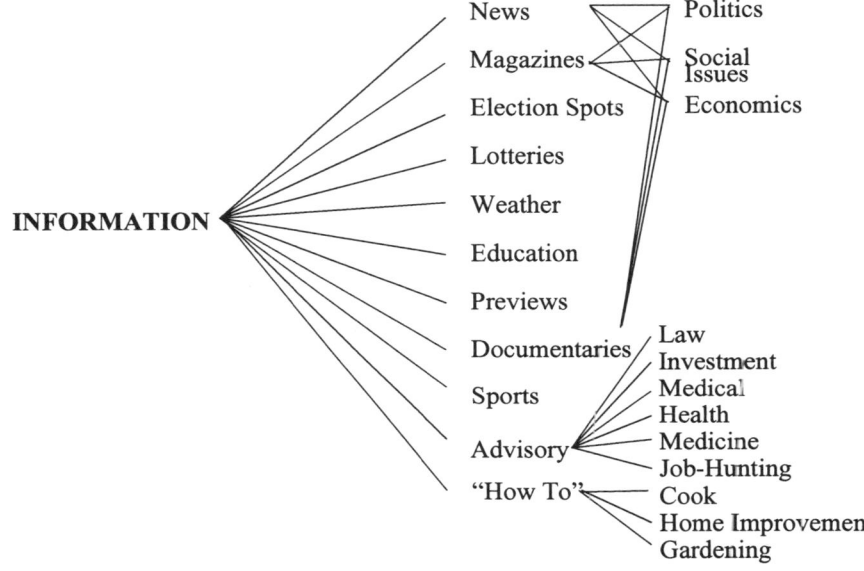

Figure 55: Subgenres of Genre "Information"

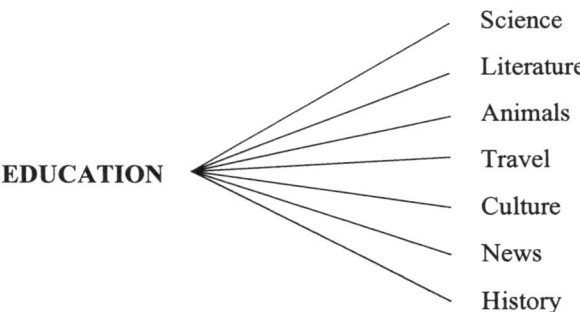

Figure 56: Subgenres of Genre "Education"

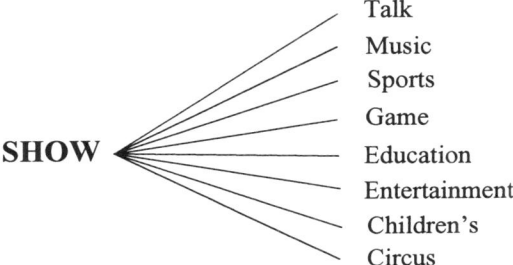

Figure 57: Subgenres of Genre "Show"

Appendix B: Personal EPG Application based on JAVA

In the following, example screen prints of the personal electronic program guide application are shown. This implementation was done in HTML and JAVA.

The personal electronic program guide is comprised of three application modules (see Figure 58): (1) subscriber management, (2) administration of user profiles, (3) generation of the personal electronic program guide.

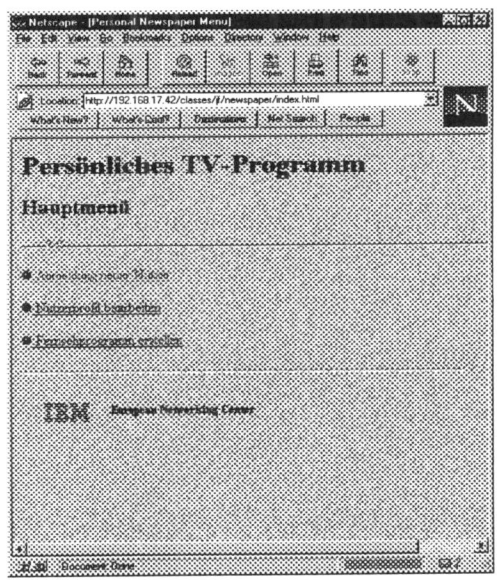

Figure 58: Main Menu

Subscriber Management

Implementation of the subscriber management modules includes functions to add and remove users and to assign logical numbers. A typical user registration procedure is shown in Figure 59. In the set-top box environment, identification and authentication is done by using personal identification cards.

Installation of User Profiles

By using the definition of preferred times of watching as an example, Figure 60 illustrates the process of installing a user profile. A user can select his preferred day and time of watching, can enter the TV channel selection menu and can save his user profile (see the screen print on in the left-hand side of Figure 60). If the profile has successfully been saved, a status message is given back to the user (see the screen print on the right-hand side of Figure 60).

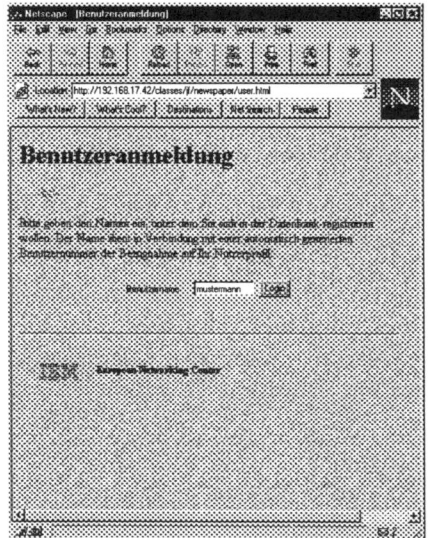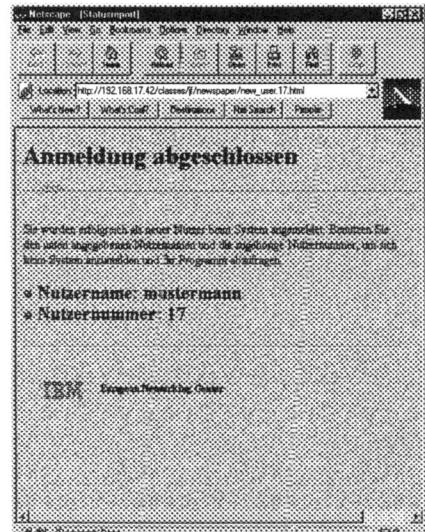

Figure 59: User Identification

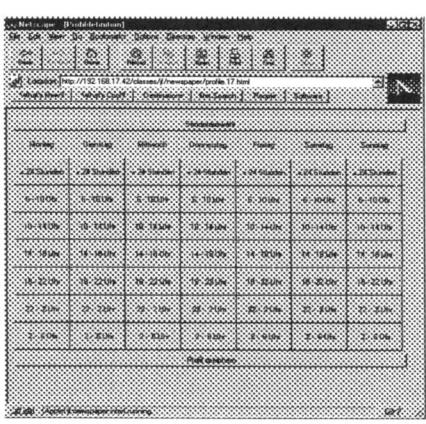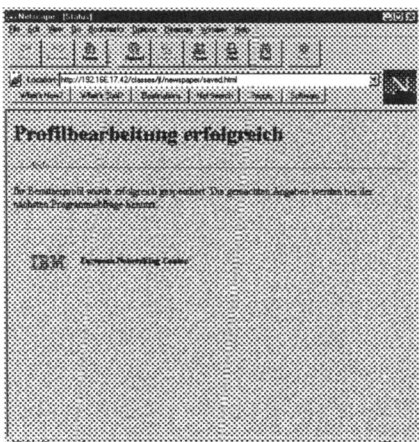

Figure 60: Definition of Preferred Times of Watching

Appendix C: The Administration Tool[15]

The Administration Tool has been developed as an administrative user interface to the personal EPG application. It is based on the intelligent media agent and is responsible for testing the intelligent agent and the personal EPG application. The Administration Tool consists of the following parts: (1) History List Interface Module, (2) Clustering Installation Interface Module, (3) Clustering Result Interface Module.

History List Interface Module

In order to test the automatic installation and update modes the channel selection, a history list interface module was implemented to place the events in the event history list of the intelligent media agent.

A user of the history list interface can select the day and the program which will then be added to the history list. The history list interface is shown in Figure 61. Each program is specified by its normal starting time and duration (e.g., starting time of the movie "1492 - Die Erorberung des Paradieses": 22.15, duration 90 min). The starting time and duration can also be set to the specific values by using slider or text boxes.

Clustering Installation Interface Module

Clustering methods are basic algorithms to derive personal preferences from the program consumed. In order to configure and test the clustering algorithm, a clustering installation interface module was implemented.

In Figure 62 an example of the clustering installation interface is shown. The sample screen print contains statistical information about the history list to be clustered (i.e., the number of events, date of last event). The following parameters of the clustering process can be configured:

- Automatic / Non-automatic clustering:
  This switch specifies whether the clustering is done automatically using a standard threshold parameter (see below), or by specifying the maximum number of clusters in the resulting profile via a box.

- Threshold:
  It specifies the minimum distance between two clusters.

---

15. The Administration Tool was designed and implemented in [Michel 97] with H.Wittig as a supervisor.

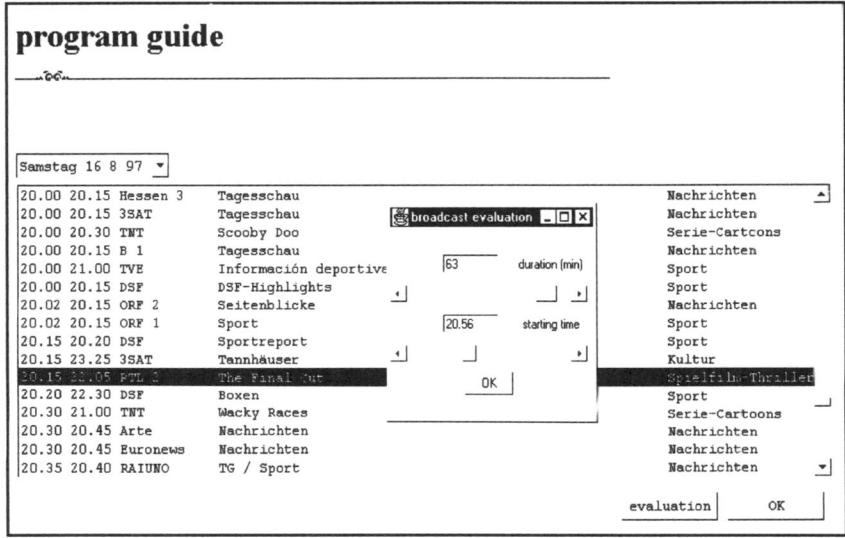

Figure 61: History List Interface Example

- Period:
  This parameter specifies the period to be used for clustering. For example, in the clustering process the entire week can be considered or clustering can be focused on specific day.
- Weight:
  These parameters influence the clustering process by defining the multiplier of the attributes time of day, genre, and priority.

Clustering Result Interface Module

The presentation of the resulting cluster structure is implemented in the clustering result interface module (see Figure 63). Traditional cluster visualization methods are provided: tables of clusters, cluster hierarchies, and dendrograms.

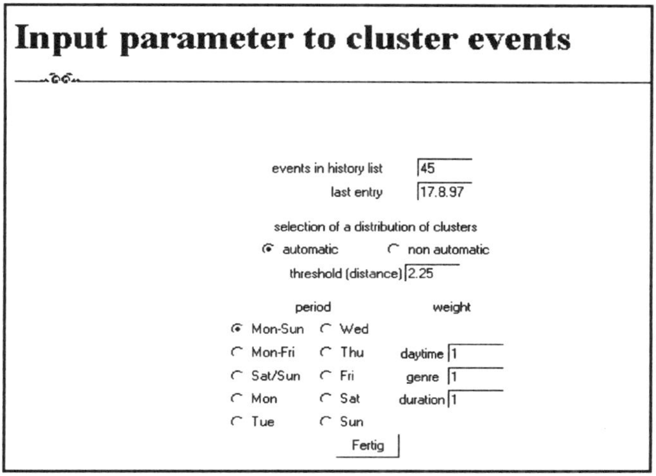

Input parameter to cluster events

events in history list   45
last entry   17.8.97

selection of a distribution of clusters
• automatic   ○ non automatic
threshold (distance) 2.25

period		weight
• Mon-Sun	○ Wed	
○ Mon-Fri	○ Thu	daytime 1
○ Sat/Sun	○ Fri	genre 1
○ Mon	○ Sat	duration 1
○ Tue	○ Sun	

Fertig

Figure 62: Clustering Installation Interface Example

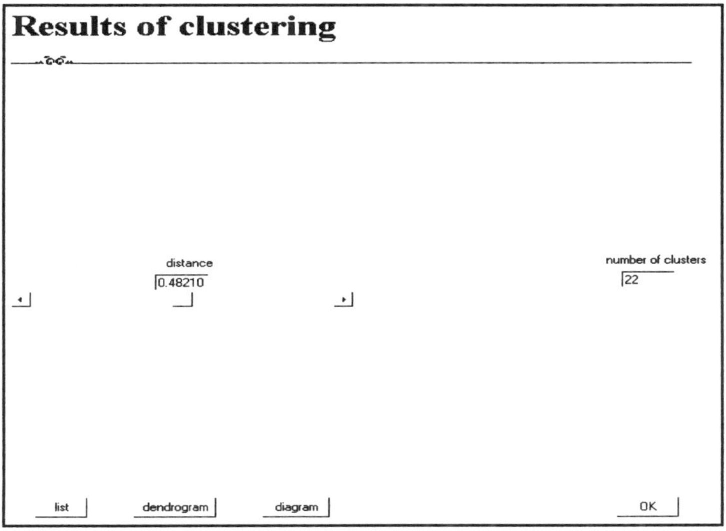

Results of clustering

distance   number of clusters
0.48210   22

list   dendrogram   diagram   OK

Figure 63: Clustering Result Interface Example

Appendix D: Sample Hit List

```
 K  -6-13    BRD                        MIO

 ----- ----01/02/95-----13:46------------
 ----------------------------------------
 *****  HITLISTE  *****

 SENDER: ARD
 DATUM : 01.01.1994-31.12.1994
 MERKM.: MIO   .K  6-13   BRD
 XV. FUSSBALL-WM USA 94

 DEUTSCHLAND - BOLIVIEN

 ARD      FR.17.06.94 20.59 109M    0.89

 XV. FUSSBALL-WM USA 94

 INTERVIEWS / MODERATION

 ARD      FR.17.06.94 21.56  10M    0.84

 DISNEY CLUB

 ARD      SA.12.03.94 16.05  84M    0.83

 DISNEY CLUB

 ARD      SA.29.01.94 16.05  83M    0.81

 XV. FUSSBALL-WM USA 94

 MODERATION / INTERVIEW

 ARD      FR.17.06.94 20.56   2M    0.76

 VERSTEHEN SIE SPASS

 ARD      SA.05.02.94 20.15 110M    0.73

 LINDENSTRASSE (425)

 ARD      SO.30.01.94 18.40  29M    0.72

 DIE DINOS

 ARD BWS FR.04.03.94 18.55  22M    0.72
```

Index

compatibility 3
complete-linkage-method 88,
  128
complexity 88, 94, 95
composer 45
construction 17, 47
Content Data Store 111
Control Agent 111, 113, 115
corporate identity 7
cost 9, 12, 22, 24, 61
cybervision 5, 25

D
data compression 1, 10
data distribution 111, 112, 115
data glove 5
DAVIC 39, 41, 43, 135
decision support system 18, 23,
  48, 97, 103, 108, 116,
  135
dendrogram 120, 164
design 12, 15, 21, 22, 52, 129,
  137
Digital Storage Media 42
digital television systems 25, 43,
  59
Digital Video Broadcasting 39
digital video broadcasting 4, 6,
  8, 61
digital video recording 6
Discrete Attribute Type 71, 81
dispatcher 45

distance function 81, 82
distance vector 64, 65, 81
distributed objects 51
distributed system 3
distribution channel 7
distribution network 4, 6, 12, 24
duration of programs 99
DVB 6, 39, 40, 41, 62, 67
dynamic approach 21, 43
dynamic presentation 39, 44

E
electronic mail 9
electronic program guide 14, 25,
  32, 41, 115, 135
electronic shopping 9, 30
exception 15, 37, 98, 103, 133,
  135
exception handling 37, 47, 98,
  101, 136
explanation module 48, 50
exploration 19, 37, 137

F
feedback 24, 37, 88, 136
feedback channel 2, 4, 6
field trial 4, 6, 8, 9, 31
filtering 10, 25, 26, 27, 28, 29,
  37, 49, 50, 51, 73, 132
Firefly 26
fixed function agent 18
functionality 2, 3, 12, 13, 15, 29,

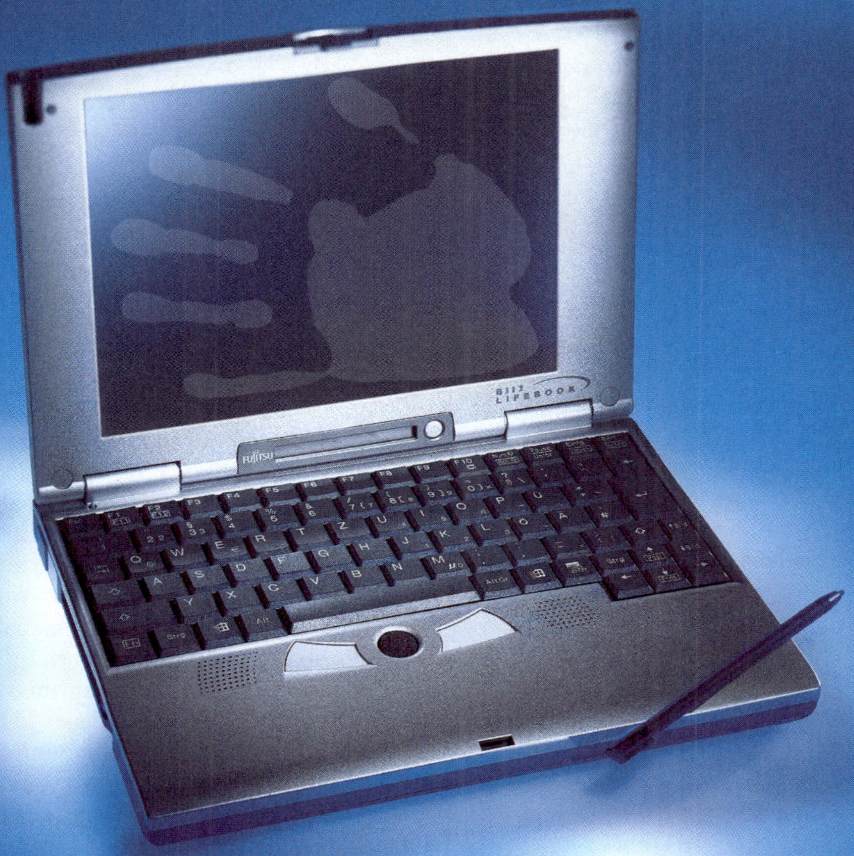

Neu!
Berühren erwünscht.

Das Biblo ist schon ein tolles Gerät: klein, leicht und jede Menge auf dem Kasten. Aber wir haben noch eins draufgesetzt. Jetzt ist das Biblo das erste Sub-Notebook mit Touchscreen. Das heißt, Sie können Ihre Eingaben direkt über den Bildschirm machen.

FUJITSU
PCs · NOTEBOOKS · SERVERS

Distribution: RFI Elektronik GmbH, Siemensring 111, 47877 Willich, Tel.: 02154/944-0

FUJITSU Lifebook B 112 Biblo ist mit einem Intel® ...ium® Prozessor mit MMX™ Technologie 233 MHz ...gestattet. 32 MB SDRAM, 3,2 GB HDD, 8,4" TFT-...lay, Li-Ion-Akku für max. 3,2 Std. Betriebszeit. Betriebssystem Windows 98 (optional Windows NT) Word 97.

FUJITSU Computer GmbH Deutschland: www.fujitsu-computer.de

Weitere Informationen über das Biblo Touchscreen sowie einen Fachhändler in Ihrer Nähe erhalten Sie unter: 0180/511 511 5* *48 Pf./Min.

..., the Intel Inside Logo, Pentium are registered trademarks and Pentium II Xeon, MMX are trademarks of Intel Corporation.

The first book about SAP®-EIS

Bernd-Ulrich Kaiser

**Corporate Information
with SAP®-EIS**

Building a Data Warehouse
and a MIS-Application with
inSight

1998. xii, 206 pp. with 44 figs. (Efficient
Business-Computing; ed. by Fedtke, Stephen)
Hardc. DM 198,00
ISBN 3-528-05674-6

Contents: Information needs and
information sources Data
warehousing inSight® for SAP®-EIS
from arcplan Building an
maintaining an Management
Information System (MIS)

The book is a real life-oriented,
professional guide to developing a
Management Information System
(MIS). The book is professional in the
sense that it adresses an MIS that
encompasses all the hierarchical
decisions-making levels within a
corporation, and it emphasizes reliable, understandable and transparent
information. The most important
demand of an MIS is an easy-to-use-
system interface, which needs to be
coupled with an information infrastructure that takes marked conditions and the company's particular
business invironment into account.
The use of a modular and flexible
system architecture is designed to
maximize the system's benefits to
cost ratio. In addition to SAP®-EIS,
the book details how to use the
inSight program (from the Duessel-
dorf-based company arcplan) to
optimize system perfomance.

vieweg

Abraham-Lincoln-Straße 46
D-65189 Wiesbaden
Fax: 0611. 78 78-400
www.vieweg.de

Stand 1.6.99
Änderungen vorbehalten.
Erhältlich im Buchhandel oder beim Verlag.